Macmillan Building and Surveying Series

Series Editor: Ivor H. Seeley
Emeritus Professor, Nottingham Trent University

Continued over

Series Standing Order (Macmillan Building and Surveying Series)

If you would like to receive future titles in this series as they are published, you can make use of our standing order facility. To place a standing order please contact your bookseller or, in case of difficulty, write to us at the address below with your name and address and the name of the series. Please state with which title you wish to begin your standing order. (If you live outside the United Kingdom we may not have the rights for your area, in which case we will forward your order to the publisher concerned.)

Customer Services Department, Macmillan Distribution Ltd
Houndmills, Basingstoke, Hampshire, RG21 2XS, England.

Sub-Contracting under the JCT Standard Forms of Building Contract

Jennie Price

MACMILLAN

First published 1994 by
MACMILLAN PRESS LTD
Houndmills, Basingstoke, Hampshire RG21 2XS
and London
Companies and representatives
throughout the world

ISBN 0–333–59657–9

A catalogue record for this book is available
from the British Library.

10   9   8   7   6   5   4   3   2   1
03 02 01 00 99 98 97 96 95 94

Copy-edited and typeset by Povey–Edmondson
Okehampton and Rochdale, England

Printed in Great Britain by Antony Rowe Ltd, Chippenham, Wiltshire

For my parents, Cliff and Beryl Price

Contents

Preface

Few people involved in the building industry would not acknowledge the importance of sub-contractors in the building process yet, with one or two honourable exceptions, textbooks on construction law touch only briefly on the standard forms of sub-contract. Having spent most of my professional life advising sub-contractors (and latterly main contractors on how to manage their sub-contractors) I have set out to redress the balance in this book.

It is intended as a guide both to the general law of contract and tort as it affects sub-contractors and to the standard forms of sub-contract for use with the major JCT forms of main contract, namely JCT 80, the Intermediate Form 1984, the JCT With Contractor's Design Form 1981 and the JCT Management Contract 1987.

Many people have helped me in the preparation of this book. They are too numerous to mention individually, but I am extremely grateful to them all. I must, however, thank Eileen McDonagh and Kay Murray, who gave me a crash course in WordPerfect when I started work on the text, unjammed the word processor on the occasions when I managed to stop it functioning altogether and corrected my numerous layout errors in the final manuscript. I must also thank my husband, Mike Hall FRICS, without whose constant advice and encouragement this book would probably never have been started, and would certainly never have been finished.

I have attempted to describe the law and the position under the standard forms as at February 1994.

JENNIE PRICE

Table of Cases

1 Basic Legal Principles

Before embarking on any analysis of sub-contracting, it is essential to have an understanding of the legal principles which govern the relationships between the parties to the building process. The two areas of law which impinge time and again on all sectors of the building industry are contract and tort (particularly the tort of negligence). This chapter therefore explains what constitutes a legally binding contract, how to decide whether a contract has come into existence, and the effect of the doctrine of privity of contract. It also sets out the basic principles of the law of negligence, and gives a legal definition of sub-contracting.

Contract

What is a contract?

A contract is an agreement or promise which is legally enforceable. There are five requirements which must be satisfied before a contract can come into existence:

(i) Intention to create a legal relationship

Such an intention will almost invariably be present if the parties are dealing with each other in a commercial context. There may, however, be special circumstances which show that the parties intend to undertake a moral rather than a legal obligation.

In the case of *Kleinwort Benson* v. *MMC* [1989] 1 WLR 379 the Court of Appeal was asked to construe a letter of comfort given by a parent company to a bank which had lent a substantial sum of money to its subsidiary. In the letter, the parent had stated:

'it is our policy to ensure that the business of [the subsidiary] is at all times in a position to meet its liabilities to you'.

The Court held that the letter did not constitute a binding contract because it was known to both sides that the parent intended to assume a moral, not a legal, obligation to repay the loan in the event of a default by its subsidiary.

Such an argument should not be regarded as an easy way out of an unattractive deal: the Court in the Kleinwort Benson case was heavily influenced by the fact that letters of comfort are generally given by parent companies when they are unwilling to provide a legally binding guarantee.

Perhaps the only circumstances when the intention to create a legal relationship can seriously be called into question in the context of sub-contracting is where an oral promise is made in a purely social environment. Otherwise, such an intention is likely to be taken for granted.

(ii) Capacity to contract

Both parties to the sub-contract must be capable of forming a legally binding relationship. There are three types of legal entity which may become involved in a sub-contract :

- a private individual, such as a sole trader;
- a partnership;
- a company.

All private individuals have the ability to form legally binding relationships, although care must be taken when dealing with minors (that is persons under eighteen years of age) and persons of unsound mind, as their capacity to contract is limited.

Smaller sub-contractors may be sole traders, which is the term used to describe a private individual using a trading name, for example Fred Jones may trade as 'Jones Heating'. When entering into a contract, Fred Jones may use his trading name, but it is Mr Jones himself who will be legally liable if there is a breach of contract by Jones Heating.

Other small sub-contractors, particularly family businesses, may operate as partnerships. Partnerships have full capacity to contract, and a partner who enters into a contract for the purposes of the partnership can act on behalf of the other partners so as to make them legally liable for the proper performance of that contract.

If either a sole trader or a partnership is successfully sued for breach of contract, the other party to the contract has recourse to all of the privately owned assets of that sole trader or of all of the partners, which is one of the main reasons why sub-contractors (and for that matter main contractors) of any size trade as limited liability companies.

Strictly speaking, a company's capacity to contract is limited by the objects specified in its articles of incorporation, as any contract which is not necessary to attain those objects is *ultra vires* (literally, beyond its powers) and therefore void. However, this does not create problems in practice for two reasons:

(i)   virtually all companies' articles are so widely drafted that they encompass all of the contracts the company enters into in the ordinary course of its business; and

(ii) under the Companies Act 1985, any transaction decided upon by the directors is, in favour of a person dealing with the company in good faith, deemed to be one into which the company has the capacity to enter.

A company can only act through its agents. In the context of entering into contracts, the directors of a company have authority to bind the company, but the extent to which other employees can do so is questionable, so it is good practice to ensure that contracts are signed by a director of the company. There are also special rules relating to the execution of deeds and contracts under seal by companies, which are dealt with below under the heading 'Formalities'.

(iii) Agreement

The agreement, or 'consensus ad idem' as it is sometimes described, is the essence of a contract. In order to decide whether there is a true agreement between the parties, the courts will look for an offer and an unconditional acceptance of that offer. These concepts are discussed below under the heading 'Formation of contracts'.

(iv) Reasonable certainty of terms

Some agreements are so vague and uncertain that they cannot give rise to binding contractual obligations. If, for example, the amount of work to be carried out by the sub-contractor is unknown, and there is neither an agreed price nor a mechanism for calculating the price, and there is no agreed time period within which the work must be carried out, it is most unlikely that a binding sub-contract will have come into existence.

However, the parties may choose to leave certain matters to be agreed after the contract has been formed – the programme and the identity of the adjudicator who is to resolve disputes on set off are common examples – but this will not prevent the sub-contract from being legally enforceable unless one of the parties has made it clear that such agreement must be reached before the contract comes into force.

(v) Consideration

Unless a contract is executed under seal or as a deed, there must be something which the law recognises as consideration if the bargain is to be legally binding.

The classic (if somewhat unhelpful) definition of consideration was given by the House of Lords in *Dunlop* v. *Selfridge* [1915] AC 847:

'an act or forbearance by one party, or the promise thereof . . . the price for which the promise of the other is bought'.

In the context of a sub-contract, the consideration provided by the main contractor is the promise to pay for the work which the sub-contractor has promised to carry out, whilst the consideration given by the sub-contractor is the promise to carry out the sub-contract works. It is important to appreciate that consideration flows in both directions: it is this reciprocity that distinguishes a contract from a gratuitous, and therefore unenforceable, promise.

Formalities

When most people think of a contract they think of a written document, but there is no legal requirement for a contract to be in writing unless it is a special type of agreement, such as a contract transferring an interest in land or a contract of guarantee.

It is therefore possible to form a sub-contract in the course of an ordinary conversation provided, of course, that the five requirements listed above have been satisfied. The problem with oral agreements is that if a dispute arises, the recollection of the parties as to what they agreed is rarely the same, which makes resolving the dispute extremely difficult. For this reason, it is good practice to confirm all oral agreements in writing.

Certain formalities are also required if the contract is to be executed under seal or as a deed. Some sub-contracts are executed 'under hand', that is the parties simply sign the document. Others are executed 'under seal' or 'as a deed'. Until July 1990, in order to execute the latter it was necessary to impress a seal on the document. As part of a package of measures designed to lighten the administrative burden on small companies, the Companies Act 1989 abolished the requirement for company seals, and introduced a system whereby a document which makes it clear on its face that it is executed as a deed, and which is signed by either two directors or one director and the company secretary, has the same effect as if it had been executed under seal. Similarly, a deed is validly executed by a private individual if it is signed by him in the presence of a witness who attests the signature and if it is delivered as a deed under the Law of Property (Miscellaneous Provisions) Act 1989.

The main difference between contracts executed under hand and those executed under seal or as a deed is that contracts under hand have a limitation period of six years, whereas contracts executed under seal or as a deed have a limitation period of twelve years. (For a full explanation of limitation periods, see Chapter 5.)

Formation of contracts

Deciding whether or not a contract has been formed, and if it has, identifying its terms, is often a difficult task. It is no problem if there is a properly executed sub-contract document, but the legal niceties are usually the last thing on everyone's minds when they are trying to get the sub-contract up and running. It is therefore very common for the sub-contract documents to be signed only after work on site has been finished. If a problem arises before then, anyone advising the parties on their legal rights will be confronted with a mass of paperwork, which is where the rules on offer and acceptance come into play.

Even if no sub-contract has been signed, if the parties have reached an agreement (and the other requirements listed above have been satisfied) a contract will have been formed. The law establishes whether an agreement has been reached by looking for an offer and an unconditional acceptance of that offer, so that is how any pre-contract negotiations must be analysed.

Invitations to tender

The first step in the process of negotiating a sub-contract is usually the issue of the main contractor's enquiry or invitation to tender. For smaller sub-contracts, this may be no more than a telephone call; for the key trades on major contracts it will involve copious documentation. This is not usually an 'offer' in legal terms but an 'invitation to treat', that is an offer to receive offers. It does not bind the person inviting the tender to contract with anyone at all – it is simply an offer to negotiate. The statement included in many invitations to tender that they do not constitute an undertaking to accept the lowest or any tender is therefore usually accurate but unnecessary, as that is the law in any event.

Tenders

The sub-contractor's tender will usually constitute his 'offer' to carry out the work, provided it is sufficiently definite and unambiguous to form the basis of a binding contract if accepted.

Sub-contractors are often asked to give main contractors budget prices or estimates for the main contractor to use in compiling his tender for the main contract works. If the sub-contractor does not wish to be bound by that budget price, he will need to take great care, as it may actually amount to an offer to carry out the work at that price. This is because the law will look at the substance rather than the form of a document when interpreting it – in other words it will look at its content rather than at what it is called. Simply describing a figure as an estimate or budget price will not be enough to ensure that it is no more than that; if it is sufficiently detailed to form the basis of a binding contract it may constitute an offer.

This risk can be minimised by including in the document a statement that the price is given for budget purposes only, and does not constitute an offer to carry out the work. It is also advisable to ensure that prices and quantities are given in round figures to ensure that the document is not as precise as a tender. In addition, an estimate is unlikely to be treated as an offer unless it is given in response to a reasonably detailed description of the works, therefore if the information supplied by the main contractor was sketchy or incomplete it is unlikely that the sub-contractor will be bound by his price.

Withdrawing tenders

In English law, an offer may be withdrawn at any time before it is accepted, regardless of any requirement or undertaking which may have been given to hold it open for acceptance for a specific period. Some invitations to tender require the sub-contractor to sign a form agreeing to hold the tender open for a specified period, but even these do not create a legally binding obligation unless the tenderer is paid to hold his tender open (which creates a collateral contract known as an option).

There is, of course, a commercial aspect of withdrawing a tender when you have promised not to do so, but if a serious error has been made in the calculation of the tender, the loss of goodwill a withdrawal will cause may be a price worth paying.

It is important to note that Scots law is fundamentally different on this point, as voluntary undertakings **are** enforceable in Scots law, and therefore no payment is necessary to bind a tenderer to hold his offer open for acceptance for a particular period if he has given an undertaking to do so.

If the sub-contractor decides to withdraw his tender, he has to do so before it is accepted. This sounds simple enough, but acceptance may have occurred without his knowledge. Under the rule in the case of *Adams* v. *Lindsell* (1818) 1 B & Ald 681, an acceptance of an offer is binding as soon as it is posted. So if the main contractor has posted an order to the sub-contractor accepting his tender before the sub-contractor has had a chance to withdraw it, he will have lost the right to do so.

The postal rule described above applies **only** to the acceptance of offers; it does not apply to withdrawals, which must be actually brought to the knowledge of the other party before they take effect. This means that if an acceptance and a withdrawal cross in the post, it is the acceptance which takes effect.

An offer does not survive for ever. If it is expressed to be open for acceptance for a period of time, it will automatically lapse on the expiry of that period. If no time is stated it will lapse after a reasonable time, which will occur when, in fairness to both parties, the person to whom the offer was made should be regarded as having rejected it by his failure to accept.

Acceptances

When an offer is unconditionally accepted, a binding contract is formed. The key word here is 'unconditional', which means that no new terms must be introduced in what purports to be an acceptance. If new terms are introduced, the document is a counter-offer which must, in its turn, be accepted before a contract can result.

Acceptance can take many forms: it can be oral or in writing or it may be inferred from the conduct of a party who starts to perform his part of the bargain, for example by starting work on site.

Although an offer can be accepted by conduct, the person making the offer may not stipulate that it can be accepted by silence. Thus a tender which states

'we will proceed on the above basis unless we hear from you to the contrary within seven days'

will not bind the recipient unless he accepts it by a positive statement or positive conduct.

With the exception of acceptances in writing sent by post which will, as stated above, take effect as soon as they are posted, acceptance must be communicated to the other party. This was confirmed in the case of *Entores Limited* v. *Miles Far East Corporation* [1955] 2 QB 327, which concerned an acceptance sent by telex. The court held that it was not until the person who made the offer knew that it had been accepted that a contract came into being. As yet, there are no cases relating to acceptance by fax, but it is thought that the *Entores* decision will be applied and therefore the contract will be created when the fax is received and brought to the knowledge of the party who made the offer, not when it is sent.

A document which would otherwise be an acceptance, such as an order on exactly the same basis as the sub-contractor's quotation, may stipulate that it is 'subject to contract'. If this is done, a contract is unlikely to come into existence until the documents have actually been executed, particularly if negotiations are continuing. However, the precise effect of these words will depend on the circumstances of the case, as a contract may be formed notwithstanding such a statement if the parties have in fact reached agreement on all of the terms.

Counter-offers

A response to an offer which introduces additional or alternative terms to those upon which the offer was made is a counter-offer. It extinguishes the original offer, and becomes the offer which must be accepted before a contract can be formed.

The negotiations leading up to a sub-contract are often a series of offers and counter-offers. A typical sequence might run like this:

- an electrical sub-contractor is invited to tender for work on the basis of a standard, unamended DOM/1 (**the invitation to treat**);
- the sub-contractor submits his tender on the same basis (**the offer**);

- the main contractor places an order which refers to the quotation, but which also contains on the reverse conditions which are inconsistent with DOM/1 (**the counter-offer**);
- the sub-contractor acknowledges receipt of the order, and starts work on site (**acceptance of counter-offer by conduct**).

In this example, the sub-contract has been concluded on the basis of the terms contained in the main contractor's order. If the sub-contractor wished to start work without accepting the amended terms, he could have done so by rejecting them before he started work on site. There is no reason why he should not start work on site **after** he has rejected them, as the terms will be accepted by conduct only if there is no express indication to the contrary. If the sub-contractor does start work on site, he will generally be entitled to payment for the work he carries out even though there is no contract between himself and the main contractor. This is considered further below under the heading 'Quantum Meruit'.

The exchange of different terms of contract is sometimes described as 'the battle of forms'. It is important to appreciate that the last shot will not always win this battle – it is not simply a matter of making sure that you have the last word if you want your terms to apply. The contract will be formed as soon as an offer by one party is unconditionally accepted by the other, and this can occur at the beginning of the battle as easily as it can at the end, particularly if acceptance is by conduct.

Letters of intent

Negotiations on the terms of the sub-contract can go on for months, so in order to get the sub-contract work started as quickly as possible main contractors often issue letters of intent to their sub-contractors. The same device may be used by an employer if the sub-contractor has been selected before the main contractor has been appointed, when the employer or his architect will often issue a letter of intent to the sub-contractor so that he can start design work or off-site fabrication. They are particularly useful for early sub-contract trades such as steelwork or piling, when an early start on site may be crucial to the timely completion of the project.

Many sub-contractors set great store by such letters, and treat them as if they were as good as, or even better than, a concluded sub-contract. Such optimism is not always justified, because the effect of a letter of intent will depend on its precise terms.

If, as the parties anticipate when the letter of intent is written, the project goes ahead and the terms of the sub-contract are eventually agreed, the letter of intent will cease to have any effect. In *Trollope & Colls* v. *Atomic Power Construction* [1962] 3 All ER 1035, the Court confirmed that the terms of the sub-contract would normally apply retrospectively to any work executed under the letter of intent.

The real problems arise when no sub-contract is ever concluded, because then the effect of the letter of intent and the obligations it imposes on the parties will be crucial. They will depend both on the terms of the letter and the circumstances in which it is written, which makes it difficult to generalise as to the rights a letter of intent will create.

A typical letter of intent will inform the sub-contractor that the main contractor intends to place an order with him for the relevant work and will request that, in the meantime, he starts work. If it does no more than that, the letter of intent will not bind the main contractor to enter into a sub-contract with its recipient, nor will it create any liability under that potential sub-contract.

This does not mean that the main contractor can escape payment for any work executed by the sub-contractor pursuant to the letter of intent; this principle was confirmed in the leading case of *British Steel Corporation* v. *Cleveland Bridge & Engineering Co. Ltd* (1981) 24 BLR 94, in which British Steel were invited to tender for the production of some cast steel nodes to be used by Cleveland Bridge in the fabrication of steelwork.

Cleveland Bridge issued a letter of intent to British Steel in which they:

- stated they intended to enter into a contract with British Steel for the nodes at the prices they had quoted;
- proposed that the contract should be on Cleveland Bridge's own standard terms;
- requested British Steel to begin work immediately pending the preparation and issue of the official form of sub-contract.

British Steel did not reply to the letter because they expected the formal order to follow shortly, but they did go ahead with the manufacture of the nodes. There were lengthy negotiations over the terms of the sub-contract, but no agreement was ever reached, and no sub-contract was signed. All of the nodes were, however, delivered.

British Steel sued Cleveland Bridge for the value of the nodes calculated on the basis of *quantum meruit* (i.e. what they were worth, or a reasonable price). Cleveland Bridge responded by counterclaiming damages for breach of contract on the grounds of late delivery and delivery out of sequence, arguing that the letter of intent had created a binding contract.

The court held that, on the facts, there was no contract between the parties because they had been unable to reach agreement on the price and other essential terms. Cleveland Bridge were therefore obliged to pay a reasonable sum for the work carried out in accordance with the request in their letter of intent. As there was no contract, British Steel were not bound to perform within a specific timescale and so Cleveland Bridge's counterclaim failed.

This case is a good example of a letter of intent which did not create any type of contract between the parties, but nevertheless obliged the author of the letter to pay a reasonable sum for work carried out under it.

In such circumstances, it is important for both parties to appreciate that there is no continuing obligation on the potential sub-contractor, and he can therefore withdraw from site (or stop doing design work) at any stage. The commercial consequences of doing so may be serious, and the sub-contractor would have to be very sure that his particular letter of intent was of the type which did not create a contract, but it is nevertheless an option which could be open to him.

One of the main risks he would take if he did stop work is that the document described by its writer as a letter of intent might, in law, actually constitute an acceptance of the sub-contractor's tender. For example, if the sub-contractor had quoted on the basis of DOM/1, the scope of the work had been reasonably well defined and the sub-contract period had been agreed, the issue of a letter of intent by the main contractor may result in a binding contract being created on the terms of DOM/1, provided that the letter did not stipulate that no sub-contract would come into existence until the documents had been executed.

Although this risk should be borne in mind, it is relatively unusual for a letter of intent to constitute an acceptance, as they are typically issued when there are still some matters to be resolved between the parties. For example, in *Kitsons Insulation Contractors* v. *Balfour Beatty Building Ltd* (1991) 8 CLD 05/04, it was held that no sub-contract had been formed by the issue of a letter of intent even though it had been counter-signed by the sub-contractor and returned before work was started. The court reached this conclusion because the matters on which no agreement had been reached, which included the method of payment, were so significant that no binding contract could come into existence until they had been resolved.

In the past, it was thought that such a result would be advantageous to the sub-contractor, particularly if he had tendered a very keen price for the work. As the *British Steel* case illustrated, in the absence of a contract, the sub-contractor will be entitled to be paid a reasonable price for the work he has carried out, and it was thought that a reasonable price would be calculated on a costs plus basis (thus allowing a reasonable addition for overheads and profit), even if commercial pressures meant that the sub-contractor's tender price did not take full account of such additions.

However, in the case of *Laserbore Ltd* v. *Morrison Biggs Wall Ltd* (1993) Building Law Monthly, vol. 10 Issue 11, it was held that 'fair and reasonable payment' meant 'a fair commercial rate for the services provided', not payment on a costs plus basis, and this principle may well be taken into account when assessing what payment is due on a *quantum meruit* basis.

Although the courts are reluctant to conclude that a letter of intent will create a contract on the basis of terms which are still being negotiated by the parties, it is quite common for them to find that the letter of intent has created some sort of ancillary contract.

For example, in *Turiff Construction Ltd* v. *Regalia Knitting Mills Ltd* (1971) 9 BLR 20, the court was asked to construe the following letter of intent:

> 'As agreed at our meeting it is the intention of Regalia to award a contract to Turiff to build a factory . . . [the letter then gave details of the work, commencement and completion dates and stated how the terms of payment were to be negotiated each month]. . .
>
> All this to be subject to obtaining agreement on the land and leases with the Corby Development Corporation, for building and bye-law consent and the site investigation. . .
>
> The whole to be subject to agreement on an acceptable contract.'

The proposed contract was to be on a design and build basis, so on receipt of this letter, Turiff began to prepare their detailed design, and entered into negotiations with Regalia's architect about a form of contract. Regalia were then taken over by another company, and the project was cancelled. Turiff sued for the costs they had incurred up to that time, relying on the letter of intent.

The judge decided that the letter as quoted was a 'true' letter of intent in the sense that it only expressed an intention to enter into a contract in the future, but created no liability in regard to that contract. However, on the facts of the case, an ancillary contract had come into being which entitled the contractor to recover the costs of the preliminary work he had undertaken pending the conclusion of a formal contract.

The words 'subject to agreement on an acceptable contract' did not prevent the ancillary contract from being formed, because they referred only to the full contract for the design and construction of the building.

The result of the *Turiff* v. *Regalia* case was therefore similar to that in *British Steel* v. *Cleveland Bridge* in that the contractor who had executed work and so incurred costs on the strength of a letter of intent was able to recover them, albeit through a different legal analysis.

It is important to note that, although neither of the letters of intent in the two cases mentioned above contained an express undertaking that the contractor should be paid for the work he had carried out if no contract resulted, that did not prevent the courts from allowing the contractors to recover payment for the work they had carried out.

Sub-contractors are often advised to obtain such an undertaking in a letter of intent before carrying out any work pursuant to it, and whilst this advice is very practical, and may help in extracting payment from a reluctant main contractor without resorting to the courts, it can backfire quite spectacularly, as was demonstrated in the case of *C. J. Sims Ltd* v. *Shaftesbury plc* (1991) 60 BLR 94.

In that case, Shaftesbury issued a document to Sims which they described as a letter of intent. The letter confirmed their appointment as contractors, and requested them to start work immediately. It also expressly stated that, in the unlikely event of the contract not proceeding, Sims would be reimbursed all their reasonable costs, including overheads. So far, so good. The problem was that the letter went on to say that those costs 'must be substantiated in full to the reasonable satisfaction of our quantity surveyor.'

Sims completed all of the work which Shaftesbury required, but no formal contract was ever concluded. They did receive some payment during the course of the works, but shortly before completion they submitted a claim for what they alleged to be their reasonable costs and their probable future costs. However, they did not 'substantiate in full' their probable future costs, so Shaftesbury argued that they should not be able to pursue their claim because proper substantiation was a condition precedent to the making of any claim at all.

This case was different to both *British Steel* v. *Cleveland Bridge* and *Turiff* v. *Regalia*, as the parties actually agreed that the letter of intent gave rise to a simple contract between them, although not on the terms of the JCT standard form on which the tender had been submitted. They therefore both relied on the terms of the letter of intent to determine their rights.

The judge found that, on the facts of the case, the requirement for substantiation was a condition precedent to payment, and so Sims were required to comply with that condition before any further sums were due to them.

Sub-contractors will, no doubt, now be much more wary of express undertakings to pay in letters of intent, particularly if they go further than a simple, unqualified undertaking to pay reasonable costs.

Quantum meruit

The phrase *quantum meruit* means, literally, 'what it is worth'. It is used to describe a claim for a reasonable sum of money for work which has been carried out in circumstances where there is no agreement as to payment. The parties may not even be in a contractual relationship, as in the case of payment for work executed under a letter of intent.

The right to be paid on a *quantum meruit* basis can arise in the following circumstances:

   (i)   where the contract does not specify a price for the work;
   (ii) where there is an express agreement to pay a reasonable sum;

(iii) where the negotiations leading up to a contract have failed, but one of the parties has carried out work on the instructions of the other, for example in a letter of intent (this is sometimes described as a quasi-contract);

(iv) where work is carried out outside the terms of a contract, for example if there is no contractual power to order additional work, but extra work is nevertheless carried out by the contractor on the employer's instructions.

The common factor in all of the above examples is that there is no existing agreement to pay an agreed sum. It is an important principle of contract law that the courts will not improve upon the bargain made by the parties. There is no inherent right to be paid a reasonable sum: if a sub-contractor has quoted a sub-economic price and that has been accepted by the main contractor, that is all the sub-contractor will be entitled to be paid. He has no implied right to be paid what his work is actually worth.

In practice, proving an entitlement to be paid on a *quantum meruit* basis is often relatively straightforward compared with the task of proving what constitutes a reasonable sum in that particular case. The courts have given very little guidance on this issue, although some basic principles have been established.

To some extent, the factors to be taken into account when assessing a reasonable sum will depend on the reason why *quantum meruit* applies. For example, if work has been performed outside a contract, the courts have held that it is wrong to have regard to the rates payable under that contract when deciding what is a reasonable sum. This prevents the party for whom the work has been carried out from taking advantage of any keen pricing in the contract: the contractor will be entitled to be paid a fair commercial rate for the work he has done.

Different rules may apply where the contract contains an express undertaking to pay a reasonable sum. For example, one of the ways in which varied work can be valued under the JCT standard forms is at 'fair rates and prices' – it is much more questionable in that case whether the paying party can have regard to the contract rates. (This point is discussed in more detail in Chapter 6.)

Another point which is often raised by those paying on a *quantum meruit* basis is that the way in which the work was carried out should have an effect on the amount payable. This argument was recently considered by the Court of Appeal, but no clear decision was reached.

In *Crown House Engineering v. Amec Projects* (1989) 48 BLR 32, Crown House carried out a substantial amount of mechanical and electrical work for Amec. No sub-contract was ever signed, and for the purposes of the case, it was assumed that no sub-contract had been formed. Amec argued that, because of the way in which Crown House had carried out the work and the time they had taken to do it, Amec had incurred extra expense and had been exposed to claims from third parties, and that these sums should be taken into account when assessing *quantum meruit*.

The judge at first instance in this case had found that, under the principle in *British Steel* v. *Cleveland Bridge*, matters such as late performance were irrelevant where there was no contract between the parties and therefore no contractual standard to be satisfied. The Court of Appeal simply noted that view, but expressly left the point open.

Until there is a clear ruling by the courts, the method and standard of performance should have no effect on the amount payable on a *quantum meruit* basis, although sub-contractors should be ready to face arguments to the contrary.

These complex rules on the assessment of *quantum meruit* will be of little comfort to a sub-contractor faced with the problem of proving what is a reasonable price for his work. A good starting point for any calculation would be the net cost of labour and materials, plus a sum to reflect overheads and a modest profit. Evidence from an independent third party such as a chartered surveyor as to the amount of work carried out and the usual rates for that type of work will also be helpful if the sum involved justifies the expense of obtaining such an opinion. If the dispute gets as far as the courts, expert evidence will usually be essential.

Recovering abortive tendering costs

The costs of preparing a tender for a sub-contract can be very substantial indeed. Tendering is a particularly costly exercise for trades such as mechanical and electrical services, where the tenderer will almost invariably be basing his price on a specification and drawings rather than on a bill of quantities, and may need to carry out a certain amount of design in order to calculate his tender price.

As a general rule, such costs are not recoverable from the person who invited the tender. The legal position was neatly summed up by the judge in *William Lacey (Hounslow) Ltd* v. *Davies* [1957] 2 All ER 712:

'If a builder is invited to tender for certain works, either in competition or otherwise, there is no implication he will be paid for the work, which is sometimes a very considerable amount of work, in arriving at his price: he undertakes that work as a gamble, and its cost is part of the overhead expenses of his business which he hopes will be met out of the profits of such contracts as are made as a result of tenders which prove to be successful.'

There are some exceptions to this principle, but it will be difficult for tenderers to prove that they fall within them. They are:

(i) Where the tender was not considered through no fault of the tenderer

This can occur where the procedure for evaluating tenders breaks down or is not properly operated.

For example, in *Blackpool & Fylde Aero Club* v. *Blackpool Borough Council* [1990] 3 All ER 25, the Council invited tenders for a contract on the basis that tenders received after the time and date specified in the invitation would not be considered. The Club put their tender into the Council's letter box before the specified time, but because the letter box was not emptied when it was supposed to be, the Council treated the Club's tender as late and refused to consider it.

In this case, the Court of Appeal departed from the usual principle that the invitation to tender was no more than an indication of a willingness to receive offers, and found it imposed a contractual obligation on the Council to consider all tenders which had been submitted in accordance with the specified procedure. The Council was in breach of that obligation, and was therefore ordered to pay damages to the Club.

This should, however, be regarded as a very limited exception to the rule that those inviting tenders have complete freedom to administer the tendering process as they see fit, as was demonstrated by the fate of a building contractor who recently attempted to rely on the decision in the Blackpool case.

In *Fairclough Building Ltd* v. *Borough Council of Port Talbot* (1992) 62 BLR 82, Fairclough's tender for a major Council project was rejected after only preliminary consideration by the Council following formal notification by a Council architect that her husband was one of Fairclough's local directors. Fairclough sued the Council for breach of contract.

The Court held that the invitation to tender had imposed on the Council an obligation honestly to consider the tenders of those they had put on their shortlist, unless there were reasonable grounds for them not doing so. On the facts of the case, they had fulfilled that obligation, and Fairclough were not entitled to any compensation.

(ii) Extra work performed at the employer's or main contractor's request

When work is scarce, contractors will often go to extraordinary lengths to get a contract. The costs they incur in doing so will normally be at their own risk, but if they carry out extra work in the course of tender negotiations at the express request of the other party to the potential contract, they may be able to recover their costs if no contract results.

This principle was demonstrated in the *William Lacey* case mentioned above, where a contractor who had submitted the lowest tender then provided additional estimates and calculations at the employer's request to enable him to negotiate funding for the project with the War Damage

Commission. When he was not eventually awarded the contract, the contractor successfully sued the employer for a reasonable sum for that additional work.

(iii) Where the employer derives a benefit

The costs of preparatory work carried out by a contractor who has been led to believe he is about to be awarded a contract may be recoverable if the contractor can prove that the employer has derived a benefit from that work. This principle was confirmed in *Marston Construction Ltd* v. *Kigass* (1989) 46 BLR 109, where Marston had tendered for a contract to rebuild Kigass' factory following serious fire damage.

Although Marston were left in no doubt by Kigass that no contract would be entered into until after the insurance money had been received, they were led to believe that they would be awarded the contract, and so carried out a considerable amount of preparatory work, including design.

In the event, the contract did not proceed because the insurers did not pay out enough to allow rebuilding in accordance with local authority requirements, and Marston claimed the costs of their additional work. The judge held that the design work had been carried out on the express request of Kigass, and that the other preparatory work had been done at their implied request. Even though there might be no actual benefit to Kigass arising out of that work, there was a potential benefit, which was sufficient to give rise to an obligation to pay for the work on a *quantum meruit* basis.

It should be stressed that all of the above cases cover exceptional circumstances: as a general rule, abortive tendering costs, or the cost of any extra work the sub-contractor does in order to improve his chances of winning the contract, will be irrecoverable.

Privity of contract

The doctrine of privity of contract is a simple but vitally important one, particularly in the context of sub-contracting. It means that only the parties to a contract can sue or be sued under it.

The parties to a sub-contract are the main contractor and the sub-contractor. Decisions of the employer or his architect may have an impact on the sub-contractor and in some forms, notably the nominated sub-contracts, the architect may even have a positive role to play, but that does not make him a party to the sub-contract. It is for this reason that the nominated sub-contracts contain a clause providing that if the sub-contractor is aggrieved by certain decisions of the architect, he may use the main contractor's name to bring proceedings against the employer under the main contract for a default by his agent, the architect.

The doctrine of privity of contract is one of the reasons why legal actions relating to building contracts usually involve so many parties. Suppose, for example, an employer discovers that his new building is not properly watertight. Extensive investigations reveal that the cause of the problem is probably bad workmanship by the sub-sub-contractor who installed the windows. The employer has no contract with that sub-sub-contractor, but he can sue the main contractor for breach of contract, who then in turn sues his sub-contractor, who in turn sues the sub-sub-contractor.

If that contractual chain breaks at any stage, for example because one of the contractors has become insolvent, or has an effective exclusion clause in his contract which means he is not responsible for the faulty workmanship, liability stops there – it cannot jump the missing link. So, in the above example, if the sub-contractor was insolvent the main contractor would have to pay damages to the employer, but would have no right of recourse against the sub-sub-contractor.

Law of tort

A tort is a civil wrong which gives rise to a right to claim damages. The most important tort in the context of building contracts is negligence; other torts, which are outside the scope of this book, include trespass, nuisance and defamation.

In order to establish a claim in negligence, the person who is making the claim (the plaintiff) must prove that the person he alleges was negligent (the defendant) owes him a duty of care and that there has been a breach of that duty which has caused a recoverable type of loss.

The concept of negligence entered English law in 1932, in the judgment of the House of Lords in *Donoghue* v. *Stevenson* [1932] AC 562. In that case, Lord Atkin formulated the principle of a duty of care in the following passage:

'The rule that you are to love your neighbour becomes, in law, you must not injure your neighbour; and the lawyer's question Who is my neighbour? receives a restrictive reply. You must take reasonable care to avoid acts or omissions you can reasonably foresee would be likely to injure your neighbour. . . [that is] . . . persons who are so closely and directly affected by my act that I ought reasonably to have them in contemplation as being so affected when I am directing my mind to the acts and omissions which are called into question.'

It is clear from this statement that a duty of care will not be imposed on everyone in all circumstances, but that an element of what lawyers describe as proximity must be present between the plaintiff and defendant.

The other crucial restriction on the right to sue for negligence is that the defendant must have suffered a recoverable type of loss. This has been an area of enormous judicial activity over the last fifteen years, when the law has swung from allowing virtually all types of loss to be recovered to the current position, which is highly restrictive.

Following decisions in cases such as *Murphy* v. *Brentwood District Council* [1990] 3 WLR 414 and *D & F Estates* v. *Church Commissioners* [1989] AC 177, to sustain a claim in negligence there must be either:

- personal injury, or
- actual physical injury to 'other property', i.e. property other than property which is the product of the negligence.

There is no right to recover damages for pure economic loss.

Applying those rules to a sub-contract in the building industry would produce the following results where a sub-contractor had failed to take reasonable care in the execution of his work:

- if the building collapsed as a result of the sub-contractor's negligence, causing physical injury to its occupants, they could recover damages from the sub-contractor;
- if the collapse damaged a neighbouring building, its owner would have a claim for damages against the sub-contractor;
- the owner of the building which collapsed would have no claim against the sub-contractor in negligence (other than in respect of the contents of the building) because he had not suffered damage to 'other property' as defined above.

The owner of the building would, however, have a right to claim damages from the sub-contractor if he had a contract with him, because economic loss can be recovered as part of damages for breach of contract.

The inability to recover economic loss in the absence of a contract is the main reason why purchasers and tenants of buildings now habitually demand collateral warranties from all those involved in the building process. The main function of such warranties is to create that vital contractual link between the people who are likely to carry out defective work or design and those who will suffer loss if they do so. (Collateral warranties are discussed in more detail in Chapter 4.)

Assuming a plaintiff alleging negligence can overcome the hurdle of proving he has suffered a recoverable type of loss, he will still have to prove that the defendant has been negligent. A sub-contractor can discharge his duty of care by exercising reasonable skill and care in the work he carries out. The standard by which his performance will be judged will be that expected of a reasonably competent sub-contractor working in the relevant field.

Where a sub-contractor is involved in design, or operates in a highly specialised field and therefore gives specialist advice, he may incur liability for negligent misstatement, which is an area of the law of negligence to which special rules apply. This is discussed in detail in Chapter 4.

Legal definition of sub-contracting

Sub-contracting is an arrangement whereby A (the main contractor) employs B (the sub-contractor) to perform part of the work A has contracted to perform for C (the employer). A remains fully liable to C for the proper performance of his contractual obligations, notwithstanding the fact that he has sub-contracted them.

Another way of describing this arrangement is 'vicarious performance'; vicarious means deputed or delegated.

Most building contracts contain an express right to sub-contract the carrying out of a part of the work, usually with the consent of the employer or his architect. If no such express right appears in the contract, it will depend on the circumstances whether such a right can be implied.

The general rule is that there is no implied right to sub-contract if there is some personal element in the obligation. Obvious examples would be a contract to write a book or to sing at a concert.

In the context of building contracts, there may be a personal element in the contractor's obligation if the contract was made with him because of his particular skill or knowledge. Certain highly specialist building trades may fall into this category, especially if there is an element of design involved, and the sub-contractor was chosen for his design skills.

Similarly, the management and coordination duties imposed by construction management and management contracts are likely to be personal and therefore not capable of vicarious performance without the employer's consent.

Distinction between assignment and novation

The crucial distinction between the concepts of assignment and novation and sub-contracting is that the relationship between the main contractor and employer remains unaffected by the sub-contracting, whereas when either an assignment or a novation occurs, it affects the relationship between the parties to the original contract because one of them, in part at least, drops out of the picture.

When an assignment occurs, a contractual right or duty is transferred from one party to another. Thus if A and B have entered into a contract under which B has agreed to pay A a sum of money, and A assigns that right to C, C takes over the right to receive that money, and could sue B if it was not paid. There is some debate as to the extent to which obligations (as opposed to rights or benefits) can be assigned, and it is certainly not possible to do so without the consent of the other party to the contract.

A novation occurs where a contract between A and B is transformed into a contract between A and C. It requires agreement between all three parties because, technically, one contract is being brought to an end, and a new one, on exactly the same terms and conditions but with one new party, is being created in its place.

Once the novation has occurred, C takes over full responsibility for everything that occurred prior to the novation, and thus is liable for any defective work executed by B, and becomes entitled to payment of any sums owed to B under the contract.

Novations occur quite frequently in building contracts, particularly where a company has become insolvent during the course of the project, and the contract is completed by another contractor. This is discussed more fully in Chapter 11.

2 Introduction to the JCT Standard Forms

Standard forms of contract

This book is primarily concerned with the standard forms of building contract published by the Joint Contracts Tribunal and with the sub-contracts which are intended for use with those forms. There are, however, a number of other standard forms of contract which are widely used in the UK construction industry, most notably the ICE forms of contract for use on civil engineering projects, and the Government Form GC/Works/1, which is used on both building and civil engineering projects by central Government departments. Although these forms and their sub-contracts share many characteristics with the JCT standard forms, there are also some fundamental differences, and therefore readers with a particular interest in the sub-contract aspects of the ICE forms or GC/Works/1 are referred to *GC/Works/1 Edition 3* by Vincent Powell-Smith (Blackwell Scientific Publications Ltd) and *Engineering Law and the ICE Contract 4th Edition* by Max Abramson.

How the JCT works

The Joint Contracts Tribunal has been producing standard forms of building contract since the 1930s. Originally a committee of architects and builders, it now comprises eleven trade associations and professional institutions (referred to as the constituent bodies) which represent all sectors of the building industry.

The current constituent bodies of the JCT are as follows:

Royal Institute of British Architects (RIBA)
Building Employers Confederation (BEC)
Royal Institution of Chartered Surveyors (RICS)
Association of County Councils (ACC)
Association of Metropolitan Authorities (AMA)
Association of District Councils (ADC)
Confederation of Associations of Specialist Engineering Contractors (CASEC)
Federation of Associations of Specialists and Sub-Contractors (FASS)
Association of Consulting Engineers (ACE)
British Property Federation (BPF)
Scottish Building Contract Committee (SBCC)

Client representation

The ACC, AMA and ADC (collectively described as the Local Authority Associations or LAAs) usually act in concert, and their policy is determined by a joint contracts panel comprising representatives of each of the three bodies.

The British Property Federation is the sole direct representative on the Tribunal of private sector clients. Although public sector clients have, traditionally, been well represented on the JCT, the Tribunal found it difficult to identify a suitable body to represent the interests of building employers in the private sector. In 1974, they invited the CBI to nominate two observers to attend JCT meetings, and although two representatives were appointed, only one attended meetings on a regular basis. That person represented the BPF, and in 1982, shortly after the publication of the BPF Manual and its accompanying contracts, the BPF were invited to become members of the JCT in their own right. The system of CBI observers lapsed at that time.

Concern is sometimes expressed by industry commentators that the views of clients are under represented on the JCT and that, as a result, it produces contracts biased in favour of the contractor. It is suggested that this is incorrect for two reasons: firstly, and most importantly, the JCT operates by consensus, which means that every constituent body has a right of veto if it does not wish to approve a particular clause or, indeed, a whole contract; secondly, the professional institutions, notably the RIBA, have traditionally represented the interests of building employers, particularly the smaller private clients.

Role of the SBCC

The SBCC is the equivalent body to the JCT in Scotland, and comprises representatives of the Scottish counterparts of the English JCT constituent bodies, with the exception of private sector clients who are represented in Scotland by the CBI and the Association of Scottish Chambers of Commerce.

How decisions are taken

The JCT operates through a series of committees which meet on a regular basis. The Tribunal itself, which must take or ratify all decisions involving a point of principle, meets six times a year.

Points of principle are also expressly referred to all of the constituent bodies, to ensure that the representatives of each of the sectional interests on the Tribunal are accurately reflecting the views of the body they represent.

In addition to the Tribunal itself, there are a number of standing committees, on which all of the constituent bodies are (or are entitled to be) represented.

The most important of these is the Drafting Sub-Committee which reviews the drafting of every single document and amendment that the Tribunal produces. Differences of view can often be resolved by drafting a clause in a particular way, and so although, strictly speaking, the Drafting Sub-Committee deals only with the way in which principles which have already been agreed are expressed, it is through the work of that Committee that many seemingly insoluble conflicts are resolved.

In addition to its standing committees, the Tribunal sets up ad hoc committees and working parties to deal with particular tasks; the revamped form of Prime Cost Contract published in 1992 was the product of such a working party.

By far the most important characteristic of the contracts published by the JCT is that they are produced by consensus. Although this can make the JCT agonisingly slow to react to changes in the law or in methods of procurement, it does mean that when a JCT standard form is produced, it takes into account the interests of all the parties to the process, and thus represents a fair balance of risk. Because consensus gives every constituent body an effective right of veto on every aspect of a new standard form or an amendment to an existing one, there is tremendous pressure to seek a compromise acceptable to all, which results in contracts which most people acknowledge are equitable, if somewhat long winded.

Scope and use of the JCT standard forms of main contract

This book is concerned with the position of sub-contractors under four of the six standard forms of main contract produced by the JCT, namely:

(1) the Standard Form of Building Contract 1980 (JCT 80)
(2) the Intermediate Form of Building Contract 1984 (IFC 84)
(3) the Standard Form With Contractor's Design 1981 (JCT 81)
(4) the Standard Form of Management Contract 1987 (the Management Contract).

Each of those forms has been written for a different type of building project, either in terms of size and complexity, or in terms of the roles of the various participants in the process. The JCT recommends in its Practice Note 20 that these forms are used in the following circumstances:

JCT 80

JCT 80 is the definitive document for use where the employer has engaged a professional consultant to design the works, and requires a contractor to carry out those works by supplying the necessary work and materials.

Although there are no express terms in JCT 80 which state that the contractor will not be responsible for design (except in relation to nominated sub-contract work), it is so obviously a contract for work and materials rather than for design, that if the employer wishes the contractor to take on a design responsibility, amendments to the standard form are advisable.

For this purpose, the JCT produces a Contractor's Designed Portion Supplement, which imposes on the contractor similar obligations to those imposed by JCT 81, in relation to that portion of the works to which the Supplement applies.

The imposition on the contractor of design responsibility in a contract which is drafted on the basis that all of the design is to be carried out by the employer's professional consultants (other than by the use of the CDP Supplement) is the source of an enormous number of problems under JCT 80, particularly in relation to design carried out by domestic sub-contractors. This is considered further in Chapter 4, which deals with the whole question of sub-contractor design.

There is no upper limit on the size of project for which JCT 80 can be used, although this traditional method of procuring buildings was often eschewed by the commercial developers responsible for the some of the biggest building projects in the 1980s, such as Canary Wharf and Broadgate, in favour of innovative systems of procurement such as construction management.

Although there is no technical reason why the form could not be used for very short and simple contracts, there are other JCT standard forms available for that purpose, and the lengthy procedures in JCT 80 would be very cumbersome on a small project.

JCT 80 is the only standard form to be produced in a number of versions: with quantities, without quantities and with approximate quantities. There are local authority and private sector versions of each of these contracts, making six versions of JCT 80 in all.

The decision whether or not to use a bill of quantities is one which the employer must make before he seeks tenders for the works. If no bill is to be produced, each tenderer will have to make his own assessment of the amount of work which will be required by examining the drawings and specification, which means that the contractor's tendering costs will be higher. There is also a risk that the contractor will make an error in his assessment, which he will be unable to remedy by claiming additional costs from the employer. In good times, therefore, employers may find that contractors' prices are higher if no bills are provided, or they may have difficulty in obtaining enough tenders for a genuine competition.

In its Practice Note 20, the JCT recommends that the use of bills of quantities is generally necessary for work estimated to have a contract value of £120 000 or more at 1987 prices, and that seeking tenders above that limit on drawings and specifications alone is not in the best interests of either the employer or the contractor. Even below that figure, bills may be desirable if the work is particularly complex, although, conversely, the JCT acknowledges that, above that figure, bills may not be required if the works are not complex, or are of a repetitive nature (such as a housing estate).

If the employer requires a lump sum price from the contractor (that is where the price quoted by the contractor is the one that he is entitled to be paid, save for adjustments to take into account variations, etc.) then either JCT 80 With Quantities or JCT 80 Without Quantities is suitable. If the employer wants work to start before detailed contract documents describing the work can be prepared, the tendering contractors will be unable to produce a lump sum price, and therefore only an indicative price, described in the JCT forms as 'a tender sum' can be given. The form suitable for use in these circumstances is either JCT 80 With Approximate Quantities, or the Management Contract, or the Prime Cost Contract.

The key characteristics of JCT 80 are therefore:

- it is intended for employer-designed works;
- there is no upper limit on the size of project for which it is suitable;
- it is too cumbersome for small, short-term contracts;
- other than the With Approximate Quantities version, it is for use where the employer requires a lump sum price.

There are two standard forms of sub-contract for use with JCT 80:

NSC/C (which recently replaced NSC/4 and NSC/4A) for use on nominated sub-contracts; *and*
DOM/1 for use on domestic sub-contracts.

The scope and use of these forms is considered in the section on sub-contracts which follows later in this Chapter.

IFC 84

The JCT published the Intermediate Form of Building Contract in 1984 in the wake of considerable criticism from the industry, particularly from the RIBA, that JCT 80 was too complex a form for the average project.

In an attempt to ensure that IFC 84 was not used for the type of project for which the Tribunal had just spent ten years developing JCT 80, certain criteria for the use of the form were endorsed on the back of the document. Although they are no more than recommendations which employers and contractors are free to ignore if they wish, it is a measure of the strength of feeling within the Tribunal that in order to circumscribe, as far as possible, the circumstances in which IFC 84 was used, they varied their normal practice of confining their recommendations to Practice Note 20.

The back cover of IFC 84 therefore states that the contract is suitable for use where the proposed building works are:

- of a simple content involving the normally recognised basic trades and skills of the industry;
- without any building services installations of a complex nature, or other specialist work of a complex nature;
- adequately specified, or specified and billed, as appropriate, prior to the invitation of tenders.

There is further guidance in Practice Note 20 as to the circumstances in which IFC 84 should be used, which stresses that the criteria on the back of the form are the most important issues to be taken into account when deciding to use IFC 84 rather than JCT 80, but that it will usually be the most suitable form where the contract period is not more than twelve months and the value of the works is not more than £280 000 at 1992 prices.

The Practice Note goes on to acknowledge that IFC 84 may be suitable for use on larger or longer contracts provided the three requirements endorsed on the back of the Form are met, but that because it is much less detailed than JCT 80, if it is used for unsuitable works, 'the equitable treatment of the parties may be prejudiced'.

The main provisions of IFC 84 and the way in which it allocates risk are generally similar to JCT 80, but as it is less than half of the length of JCT 80, a lot of the procedural elements found in that document have been omitted.

Like JCT 80, it is intended for use where that employer's professional consultants have designed the works, and therefore it is a contract for work and materials rather than for design. There is no Contractor's Designed Portion Supplement specifically drafted for use with IFC 84, and therefore the only way in which an employer can obtain design from the contracting side without amending the document is through the use of named sub-contractors. This is discussed in more detail in Chapter 4 on Design by Sub-Contractors.

IFC 84 can be used in conjunction with either a bill of quantities or a specification and drawings. There is a single version of IFC 84 for use with or without bills and in both the public and private sectors.

It is a lump sum contract and so there is no opportunity for the submission of a tender sum and remeasurement on completion.

The key characteristics of IFC 84 are therefore:

- it is suitable for use in broadly the same circumstances as JCT 80 *but*
- it is intended for simpler contracts of short duration and a value not exceeding £280 000 at 1987 prices, *and*
- it is invariably a lump sum contract.

There are two standard forms of sub-contract suitable for use with IFC 84, namely:

NAM/SC for use on named sub-contracts; *and*
IN/SC for use on domestic sub-contracts.

The scope and use of these forms is considered in the section on sub-contracts which follows later in this chapter.

JCT 81

In 1981, the JCT responded to the increasing popularity of design and build, sometimes described as 'package deal contracting', by publishing a version of its standard form 'With Contractor's Design'.

This contract is drafted on the basis that the contractor will design either the whole of the works or, if they have been partially designed when tenders are invited, that the contractor will complete the outstanding design.

The contract requires the employer to state his requirements for the building, which can be anything from a general description of the accommodation required up to a full scheme design of the works – or even, in theory at least – a detailed design.

In response, the contractor submits his proposals for carrying out the works, together with a lump sum price and a contract sum analysis which will be used for valuing changes in the employer's requirements (i.e. variations).

An important characteristic of JCT 81, which distinguishes it from all of the other contracts produced by the Tribunal, is that it does not require the employer to appoint an architect or contract administrator to administer the contract. Although the employer must appoint an agent who acts on his behalf, that person does not act as a certifier and could be an employee of the employer's own company rather than a professional consultant.

If there is to be a substantial element of contractor-input into the design of the works (other than by named and nominated sub-contractors appointed under JCT 80 and IFC 84 respectively), the employer will have to decide whether to use JCT 81 or JCT 80 as modified by the Contractor's Designed Portion Supplement.

Practice Note 20 is somewhat oblique on this point, but perhaps the best way to describe what the Tribunal appears to intend is that if the employer intends his architect to retain control over the design as a whole, and to pass design responsibility in relation to only a section of the works (e.g. the heating and air conditioning systems) to the contractor then the CDP Supplement will be appropriate. However, if the contractor is to take over responsibility for completing the design of the whole of the works, no matter how far advanced the design may be at that time, JCT 81 will be the appropriate contract.

JCT 81 is particularly popular with property developers, perhaps because it is the JCT standard form which most closely resembles the contracts produced by the Association of Consultant Architects for use with the BPF system. An increasingly common practice amongst such clients is to employ an architect to produce an outline or scheme design, and then to employ a contractor on the basis of JCT 81 to complete the design and build the building, but also to require the contractor to accept a novation of the architect's conditions of engagement and to employ him to complete the design.

This arrangement is almost invariably accompanied by amendments to the standard form requiring the contractor to take responsibility for the design of the whole of the works, including that carried out by the employer's designer prior to the novation. Thus the employer retains control over the aesthetically important stages of the design process but obtains the single-point responsibility that is one of the major benefits of design and build. There are, however, dangers for both the contractor and sub-contractor in this process, which are explored in Chapter 4.

The key characteristics of JCT 81 are therefore:

- it is for use where the contractor is to design the whole or a substantial part of the works (although in theory it may be used where the contractor has very little input into the design);
- it is a lump sum contract;
- the employer is not required to appoint an architect or contract administrator, although he must name someone to act as his agent for the purposes of the contract.

There is only one standard form of sub-contract suitable for use with JCT 81, namely DOM/2. It is described in more detail in the section on the scope and use of sub-contracts.

The Management Contract 1987

The Tribunal published its standard form of management contract in 1987, well after management contracting had become popular in the UK. This was partly because, traditionally, the JCT has followed trends in methods of procurement rather than initiated them, and also because of the need to obtain consensus on the content of the Management Contract. This was a difficult task in itself, not helped by the fact that while the document was being negotiated, there was serious conflict between some constituent bodies on other issues relating to amendments to JCT 80. However, these were eventually resolved, and the JCT form of management contract was published.

It is a relatively 'pure' version of management contracting, which is deliberately low risk for the management contractor. It is his job to manage the process, whilst all of the work on site is carried out by works contractors.

In its Practice Note MC/1, the Tribunal recommends that suitable conditions for the use of the Management Contract would be where:

- the employer wishes the design to be carried out by an independent architect and design team;
- there is a need for early completion;
- the project is fairly large;
- the project requirements are complex;
- although he requires early completion, the employer wants the maximum possible competition in respect of the price for the building works.

The Management Contract is not a lump sum contract. The employer's quantity surveyor prepares a contract cost plan which is an indication of the price that the employer will pay for the building (excluding the management contractor's fee). The management contractor is entitled to be paid the actual cost of building, described as the prime cost (which is defined in great detail in one of the schedules to the contract). In addition to that prime cost, the employer pays the management contractor a fee for his management services, which can be either a lump sum or a percentage of the contract cost plan total.

Uniquely, the Management Contract provides for the project period to be divided into two phases: the pre-construction period, during which the programme is prepared, early works contractors are selected and appointed and, most importantly of all, the management contractor advises on the practical aspects of design; and the construction period, which is when work is carried out on site.

The Management Contract also contains unique provisions relating to the management contractor's responsibility for any default by the works contractors whom he employs to carry out the work on site. Although he is prima facie liable to the employer in the event of any default by a works contractor regarding defective work or materials or delay, he is relieved of the consequences of that default under a complex set of 'relief provisions', which are described generally in Chapter 11 on determination, in Chapter 5 in relation to defects, and in Chapter 8 in relation to delay.

The key characteristics of the Management Contract are therefore as follows:

- it is low risk for the management contractor;
- it is suitable for use for large and complex projects where early completion is required;
- it is a prime cost rather than a lump sum contract.

There is a single standard form of works contract (it is not strictly correct to describe it as a sub-contract) for use with the Management Contract, namely Works Contract/1, which incorporates by reference the standard conditions set out in Works Contract/2. This contract is discussed in more detail in the section on the scope and use of standard forms of sub-contract.

Other JCT standard forms

In addition to the forms described in detail above, the JCT currently publishes two other standard forms of main contract which are outside the scope of this book. The Agreement for Minor Building Works 1980, which is intended for very low value contracts (£70 000 at 1992 prices) of a simple nature. Although widely used, particularly for domestic refurbishment work, it contains no detailed provisions relating to sub-contractors and there is no standard form of sub-contract suitable for use with it.

The Standard Form of Prime Cost Contract, which was completely revised and updated in 1992, is used relatively rarely and is intended for circumstances where there is a need for work to start before the extent of the works has been ascertained, for example the repair of fire damage. Although there is a special version of the nominated sub-contract for use with the prime cost form (NSC/PCC), it is identical to NSC/C save for the references to the main form which relate to the Prime Cost Form rather than to JCT 80.

Scope and use of standard sub-contracts

Sub-contracts for use with JCT 80

There are two standard forms of sub-contract for use with JCT 80, but only one of them, the nominated form, is produced by the Tribunal itself. It is only comparatively recently that the JCT began to publish any standard forms of sub-contract at all; many people do not realise that the old 'green' and 'blue' forms of sub-contract for use with the 1963 Edition were published not by the Tribunal but by bodies representing the main contractors and sub-contractors.

JCT 80 was the first JCT standard form to be accompanied by its own set of JCT sub-contract documents, which were for use on nominated sub-contracts only. Before Amendment 10 to JCT 80 was issued in March 1991, there were two methods of nomination, the basic method and the alternative method, each with its own set of JCT sub-contract documents.

Where a sub-contractor was nominated under the basic method, the use of the following documents was mandatory under clause 35 of JCT 80:

NSC/1: form of tender
NSC/2: employer/sub-contractor agreement
NSC/3: nomination instruction
NSC/4: form of sub-contract

Where a sub-contractor was nominated using the alternative method, standard forms were available but, with the exception of the sub-contract itself, their use was optional. The forms were as follows:

NSC/1a: form of tender
NSC/2a: employer/sub-contractor agreement
NSC/3a: nomination instruction
NSC/4a: form of sub-contract

Amendment 10 to JCT 80 merged the two systems of nomination into a single procedure, and therefore a new set of nominated sub-contract documents had to be produced. The use of the following documents is therefore mandatory under Amendment 10 wherever a nominated sub-contractor is appointed:

NSC/T: form of tender
NSC/A: articles of agreement
NSC/C: standard sub-contract conditions, which are incorporated by reference into NSC/A
NSC/W: employer/sub-contractor agreement
NSC/N: nomination instruction

NSC/T is quite different in layout to the standard forms of tender that it replaces, although the information required from the parties is essentially the same. NSC/A and NSC/C, taken together, replace the sub-contracts NSC/4 and NSC/4a. The content of the sub-contract was not changed by Amendment 10, but the conditions were reorganised into a section headed format which means that all the clause numbers were changed.

Wherever reference is made in this book to a clause in the standard form of nominated sub-contract, the clause number in NSC/C will be given.

The nomination procedure and the advantages and disadvantages of nominating sub-contractors are discussed in Chapter 3.

The other form of sub-contract for use with JCT 80 is DOM/1, the standard form for use on non-nominated, or domestic, sub-contracts. The use of DOM/1 is not mandatory in any way: it is for the main contractor and domestic sub-contractor to decide what terms will govern their relationship. (The use of non-standard sub-contracts is considered below.)

DOM/1 is not published by the JCT, but by the representative bodies of main contractors and sub-contractors, namely BEC, FASS, CASEC and FBSC. (FBSC is a sector of the BEC which represents traditional building sub-contract trades such as stonemasons and painters.)

The form is published in two parts. The recitals, articles of agreement and appendix are contained in one booklet, and they incorporate by reference the standard conditions, which are contained in a separate booklet. This approach was adopted to ease the administrative burden on the parties; the intention is that main contractors and sub-contractors should need to purchase only one copy of the conditions for reference purposes.

Sub-contracts for use with IFC 84

IFC 84 provides for the appointment of two types of sub-contractor: named and domestic.

The JCT produces a standard form of named sub-contract, NAM/SC, and a standard form of tender, NAM/T. Under clause 3 of IFC 84, both of these documents must be used on every named sub-contract. In addition, RIBA and CASEC have published an employer/sub-contractor agreement, known as ESA/1.

The naming procedure and the function of the various documents is described in detail in Chapter 3.

A standard form of domestic sub-contract for use on domestic sub-contracts let under IFC 84 has been published by BEC, FASS, CASEC and FBSC. This form, known as IN/SC, is almost identical to DOM/1. Like DOM/1, it is published in two parts, with the recitals, articles of agreement and the appendix in one booklet, and the conditions in another.

Sub-contract for use with JCT 81

The Tribunal itself has not published any standard forms of sub-contract for use with the JCT design and build form because all sub-contractors appointed under that form are essentially domestic, although there is a limited provision for naming in the supplementary provisions.

However, BEC, FASS, CASEC and FBSC have issued a standard form of sub-contract known as DOM/2. This is a awkward document to use, because it operates by providing for certain amendments to be made to the DOM/1 conditions.

If you buy a copy of DOM/2, you are supplied with a slim booklet containing a set of recitals, articles of agreement, appendix and a 'Schedule of Amendments to the Conditions', which are the amendments to the DOM/1 conditions needed to make them suitable for use as a sub-contract under JCT 81. This is fine so far as it goes, but if one needs to refer to the DOM/2 conditions it can be almost impossible to interpret the clauses without laboriously inserting all of the amendments made in the schedule into a copy of the DOM/1 conditions.

It is understood that the publishers of DOM/2 intend to produce a substantive version of the DOM/2 conditions so that this process will no longer be necessary, but at the time of writing DOM/2 remains a difficult contract to use.

It is a commonly held view that DOM/2 is the correct form of sub-contract to use wherever the sub-contractor has an input into the design of the building. This is completely incorrect. DOM/2 was drafted as a sub-contract for use with JCT 81, and although several of its provisions undoubtedly deal with design, others step down procedures which are unique to JCT 81 but are unrelated to design, such as special procedures regarding final payment.

Using DOM/2 as the sub-contract when JCT 80 or some other standard form is the main contract can lead to disaster, because the main contract and sub-contract will not be 'back to back'. If a domestic sub-contractor appointed under JCT 80 is contributing to the design (a frequent occurrence but one for which JCT 80 does not expressly cater), then the best course is for the main contractor to make appropriate amendments to DOM/1, assuming of course that he wishes to use a standard form of sub-contract at all. This issue is discussed in more detail in Chapter 4.

Works contract for use with the Management Contract

The JCT Management Contract describes the contractors who carry out the work on site as works contractors rather than sub-contractors, and therefore that is the term used throughout this book.

The JCT has produced a standard set of documentation for use on works contracts, which is as follows:

Works Contract/1: the tender form, which also includes the recitals and the articles of agreement
Works Contract/2: the conditions, which are incorporated by reference into the articles
Works Contract/3: the employer/works contractor agreement

All of these documents must be used, in their standard, unamended form, unless the employer and management contractor otherwise agree.

The process of selecting works contractors under the Management Contract is discussed in detail in Chapter 3.

Use of non-standard sub-contracts

Although there are many standard forms of sub-contract available, some of which have been listed in this chapter, the parties generally have the right to agree whatever terms they choose to regulate their contractual relationship. The only exceptions to this rule are nominated sub-contracts under JCT 80, named sub-contracts under IFC 84 and works contracts under the Management Contract, where the main contract requires the use of a standard form of sub-contract.

Although that might sound like a lot of exceptions, the vast majority of sub-contractors are domestic, and therefore the parties have a free hand, subject to the restrictions imposed by the general law.

Freedom of contract is a basic tenet of English law, but in the last twenty years this principle has been eroded both by statute and decisions of the courts. The most important restrictions are those imposed by the Unfair Contract Terms Act 1977. Although this Act does not outlaw unfair contract terms generally, it limits the extent to which a party can exclude or restrict his liability in his own standard terms.

Section 3 of the Act provides that, where a party contracts on his own written standard terms of business, he may not exclude or restrict his liability for breach of contract, or claim to be entitled to render a contractual performance substantially different from that reasonably expected of him, except to the extent that the relevant term satisfies the requirement of reasonableness.

Whether or not a particular term is reasonable will depend on a number of factors, including the relative bargaining power of the parties, so smaller sub-contractors do have some protection against one-sided sub-contracts. However, the Act does not restrict in any way whatsoever the positive obligations which may be imposed on the other party (i.e. the sub-contractor), and therefore that protection is limited. It would also be expensive to enforce, as it is difficult to prove that a particular term is unreasonable under the Act without actually going to court.

There is also some comfort for the victims of unfair sub-contracts in the judgement of the Court of Appeal in the case of *Interfoto Picture Library v. Stiletto Visual Programmes Ltd* [1989 QB 433]. In that case, it was held that if one condition in a set of printed conditions was particularly onerous or unusual, the party seeking to enforce that condition must show that it had been brought to the attention of the other party to the contract at the time the contract was made. This will be of little help where a clause is common, such as a pay-when-paid clause, but it could be very useful in resisting the more ingenious non-standard clauses.

Most of the larger main contractors have their own forms of sub-contract, which may be a variant on a standard document such as DOM/1, or a completely original set of terms and conditions.

Such forms are the subject of heated debate between main contractors and sub-contractors. Sub-contractors claim that they create an environment of 'order above and chaos below' – that is balanced, equitable terms and conditions at main contract level and onerous, one-sided conditions at sub-contract level. To an extent this is true, but it is also the case that the favourability of the contract terms to one party or the other is a product of market forces, and therefore if the main contractor is in the dominant position, which is usually the case in a recession, he will be able to dictate the terms of the sub-contract. It is also instructive to consider the terms which many of the strongest critics of main contractors impose on their own sub-sub-contractors; they are often as onerous, if not more onerous, than main contractors' ad hoc conditions.

There are, however, certain trades that traditionally enjoy a strong bargaining position owing to a limited number of competitors, for example liftmakers. They can often secure much better conditions than those that operate in bigger, less specialist markets, such as plumbers or painters.

Although there are undoubted advantages in the use of standard forms, including familiarity, lower tendering costs and, in theory at least, lower prices, it is unlikely they will ever be used on a universal basis. However, the standard forms are the basis on which most major sub-contracts are let, and they therefore form the basis of the comments in this book on the position of domestic sub-contractors. Where amendments are commonly made to standard clauses, for example the inclusion of pay-when-paid clauses, they are also commented on in the relevant chapter.

3   Selecting Sub-Contractors

All of the JCT standard forms (with the exception of the Management Contract) allow for ordinary domestic sub-contracting, which gives the main contractor complete freedom of choice when selecting his sub-contractors. But they differ quite dramatically in the extent to which they allow the employer to select sub-contractors, and in the contractual consequences of doing so.

The question of who has the right to select sub-contractors, and what the contractual effect of exercising that right should be, has been one of the most hotly debated subjects in the Joint Contracts Tribunal in recent years. Most employers want the right to choose who should carry out key elements of the work such as mechanical and electrical services, but they are reluctant to accept that, if they do so, the main contractor's responsibility for that sub-contractor's performance should be limited.

Historically, any employer who wanted to select a sub-contractor used nomination, but that system is now much less popular than it once was, particularly in the South of England where nomination is almost unheard of on private sector contracts. Although major private sector employers have experimented with other methods of employer selection such as naming, the only JCT standard form to cater for it in any detail is IFC 84. As a result, the clauses relating to sub-contracting are among the most frequently amended in JCT 80.

This chapter describes the various ways in which a sub-contractor can be chosen, examines the standard tendering procedures which may be imposed on the parties by the main contract, and sets out the advantages and disadvantages of the different methods of selection.

Nomination

Nomination is the term used to describe the selection of sub-contractors by the employer under clause 35 of JCT 80. It is sometimes used colloquially to describe any employer selected sub-contractor under any standard form, but this is both confusing and inaccurate. Only JCT 80 and the JCT Prime Cost Contract provide for the appointment of nominated sub-contractors, and far-reaching consequences flow from the use of the nomination process which, as will become apparent in this chapter, do not apply to other methods of procuring sub-contract work.

Key features of nomination

The key features of the nomination system are:

- the sub-contractor is selected by the employer and nominated to the main contractor by the architect;
- the use of a standard tendering procedure set out in clause 35 of JCT 80 is mandatory;
- the main contractor must employ the sub-contractor on the terms of the standard form of sub-contract, NSC/C;
- under clause 35.21 of JCT 80, the main contractor is not liable for any design carried out by nominated sub-contractors;
- the nominated sub-contract work is the subject of either a prime cost or provisional sum, and is paid for by the employer at the nominated sub-contractor's rates and prices, plus an allowance for the main contractor's profit and attendances;
- the architect participates in the administration of the sub-contract and makes decisions on key issues such as payment, practical completion and extensions of time;
- the main contractor is entitled to an extension of time for delays caused by the nominated sub-contractor;
- the employer bears the bulk of the additional costs incurred in the event of the determination of the nominated sub-contractor's employment, whether it is due to a default by the sub-contractor or his insolvency.

The nomination procedure

Until 1991 there were two methods of nomination: basic and alternative. The standard documents used where each of these methods applied are listed in Chapter 2 for the sake of completeness, but as they are no longer in general use, they are not covered in detail in this chapter. It is, however, interesting to note that the basic method involved the use of a rather complicated standard form of tender which was much criticised in the industry, particularly by architects, whilst the alternative method did not require the use of any standard tendering procedure at all. As a result, the alternative method was the more popular of the two.

Towards the end of the 1980s, the JCT set up a working party to review a number of matters relating to sub-contractors, which concluded that the nomination procedure could and should be simplified. That simplification took place in Amendment 10 to JCT 80, published in March 1991, which replaced the basic and alternative methods with a single system described as 'the 1991 procedure'.

It is important to appreciate that, although Amendment 10 completely revised the procedural aspects of clause 35, it did not change the allocation of risk between the parties as a result of nomination. Because much of the animosity to nomination is due to the liabilities the employer is forced to assume when he decides to nominate, it is debatable whether the changes will rehabilitate nomination to the extent many expect.

The 1991 procedure works as follows:

1. The architect completes the invitation to tender in part 1 of the standard form of tender, NSC/T. He also completes the relevant section of the employer/nominated sub-contractor agreement, NSC/W, and sends both documents, together with the drawings/specification/bill of quantities which describe the work, to those companies the employer wishes to invite to tender.
2. The tenderers complete the standard form of tender (part 2 of NSC/T) and the relevant sections of the warranty, NSC/W, and return both documents to the architect.
3. The architect selects the sub-contractor, and arranges for the employer to sign the selected sub-contractor's tender to confirm his approval of it, and to execute the warranty, NSC/W.
4. The architect nominates the sub-contractor to the main contractor using the standard nomination instruction, NSC/N. He also sends the main contractor a copy of the invitation to tender and the successful sub-contractor's tender (i.e. Parts 1 and 2 of NSC/T), the drawings/specification/bill of quantities on which the tender was based, and the executed warranty.
5. On receipt of those documents, the main contractor has ten days within which he may exercise his right of reasonable objection to the selected tenderer. If he does not exercise that right, he must reach agreement with the selected tenderer on the 'Particular Conditions' set out in Part 3 of NSC/T. These include the sub-contract period, insurance details and the identity of the adjudicator and trustee stakeholder.
6. When agreement on the Particular Conditions has been reached, the contractor and sub-contractor sign Part 3 of NSC/T, and execute the sub-contract articles of agreement, NSC/A, which incorporate by reference the sub-contract conditions NSC/C.
7. The main contractor sends a copy of the signed NSC/T Part 3 and the executed NSC/A to the architect for his records.

The purpose of executing the warranty NSC/W at a relatively early stage is to create a binding contract between the employer and the successful tenderer so that the architect can instruct the sub-contractor to proceed with design work, the purchase of goods and materials or the fabrication of components before a sub-contract is executed. The employer will be obliged to pay for this work under NSC/W if for any reason the nomination process breaks down and no sub-contract is concluded.

Problems often arise at Stage 5, when the main contractor tries to agree the Particular Conditions with the sub-contractor. Under Clause 35.8 of JCT 80, agreement must be reached within ten working days of the main contractor's receipt of NSC/N. If this cannot be achieved, the main contractor must give the architect a written notice stating either the date by which he expects agreement to be reached or, if lack of time is not the problem, the other matters which are preventing agreement.

If the parties simply require more time, the architect may, having consulted with the contractor, fix a later date by which the sub-contract must be entered into. However, if this prolongation of the negotiations causes a delay to the completion of the main contract works, the main contractor will be entitled neither to an extension of time nor to loss and expense.

If other matters are preventing agreement, the architect must decide whether, in his opinion, they justify a failure to agree. If he decides they do not, he can simply instruct the contractor to reach agreement with the sub-contractor and execute NSC/A.

Clause 35 does not impose any special sanctions for failure to comply with that instruction, so it would seem that the remedies available to the employer if the main contractor is still unable to reach agreement with the sub-contractor will be the same as for non-compliance with any other instruction, namely employing and paying others to execute the work necessary to give effect to the instruction (which would probably allow the employer to employ the proposed sub-contractor direct) and recovering any additional costs of doing so from the main contractor.

Where the architect is of the view that there are genuine reasons for the failure to agree, he is obliged to issue further instructions under clause 35.9.2 of JCT 80 either:

- to facilitate agreement (for example if the problem is related to agreeing the sub-contract period, the architect may be able to allow the main contractor to accommodate the sub-contractor's requirements by amending the main contract period); *or*
- to omit the work (in which case the employer could employ the proposed nominated sub-contractor direct); *or*
- to nominate another sub-contractor.

Any such further instructions would entitle the main contractor both to an extension of time and to loss and expense.

Advantages and disadvantages of nomination

It is difficult to list the advantages and disadvantages of any method of procuring sub-contract work in the abstract, as what will be a benefit to one participant in the process might be a drawback to another. The advantages and disadvantages of nomination for each of the key players – the employer, the main contractor and the sub-contractor – have therefore been identified separately.

The **advantages** of nomination for **the employer** are:

(i)   the ability to decide who should carry out the sub-contract work;
(ii) the ability to harness the sub-contractor's design expertise through appointment at an early stage, possibly before the main contractor has been appointed. This is particularly valuable in relation to trades such as mechanical and electrical services which, traditionally, make a major contribution to the detailed design of their work;
(iii) a direct contractual link with the sub-contractor, through the warranty NSC/W, which covers matters such as design and selection of materials by the sub-contractor.

The **disadvantages** of nomination for the **employer** are:

(i)   despite simplification by Amendment 10, the standard procedure prescribed by JCT 80 remains somewhat cumbersome and can be difficult to operate;
(ii)   no rights against the main contractor in the event of defective design by the sub-contractor;
(iii) if the progress of the main contract works is delayed by a nominated sub-contractor, the main contractor will be entitled to an extension of time under clause 25.4.7 of JCT 80 and will not therefore be required to pay liquidated damages in respect of that delay – although the employer will have the right to recover damages for delay from the sub-contractor under the warranty NSC/W, this can be difficult to implement and in any event will be of no use if the sub-contractor is insolvent;
(iv)   because the main contractor is not responsible for delay by nominated sub-contractors, there may be a lack of management of the progress of the sub-contractor's work on site;
(v)   blurred lines of communication and management because the sub-contractor will be answerable to the architect on some issues and to the main contractor on others;
(vi)   the employer bears all extra costs arising out of a determination of the nominated sub-contractor's employment on the grounds of either default or insolvency, except for any delay caused by the architect taking a reasonable time to renominate and for the costs of the rectification of defects in work executed by the departed nominated sub-contractor (although that rectification work must be included in the work to be carried out by the new nominated sub-contractor).

The **advantages** of nomination for the **main contractor** are:

(i)   he will not be responsible for delay caused by nominated sub-contractors, although in order to obtain an extension of time on this ground he will have to show that he has taken all practicable steps to avoid or reduce that delay;

(ii) there is a clear exclusion of any liability on the main contractor in respect of any design carried out by nominated sub-contractors;

(iii) he bears only very limited risks in the event of the determination of the sub-contractor's employment, namely for any delay caused by a reasonable time taken by the architect to renominate and the costs of rectifying defects in work executed by the sub-contractor whose employment has been determined.

The **disadvantages** of nomination for the **main contractor** are:

(i)   he loses the right to choose who carries out that part of the works – most main contractors are reluctant to try to exercise their right of reasonable objection unless there are very clear grounds, such as where they are already involved in major litigation with the proposed sub-contractor;

(ii) he is forced to employ the sub-contractor on the basis of a standard form of sub-contract, and so is unable to impose his own terms and conditions which may be more commercially advantageous to him;

(iii) nominated sub-contractors can be difficult to manage and control because of their direct links with the professional team and the division of contractual responsibility for their performance, which is owed to the employer in respect of design and progress and to the main contractor in respect of quality of work and materials.

The **advantages** of nomination for the **sub-contractor** are:

(i)   a standard tendering procedure which both ensures quality and consistency of information and prevents Dutch auctioning (i.e. the practice of negotiating reductions after competitive tenders have been submitted by, for example, telling the second lowest tenderer he is £10 000 higher than the lowest and inviting him to reduce his price);

(ii) a standard form of sub-contract which represents a fair balance of risk between main contractor and sub-contractor;

(iii) involvement of the architect in matters such as determining extensions of time and loss and expense due to the sub-contractor;

(iv) payment on the basis of architect's certificates;

(v)   contractual acknowledgement of and, in theory at least, payment for the sub-contractor's input into the design process.

There are no obvious disadvantages for the sub-contractor in nomination.

Circumventing the nomination procedure

Because of the risks imposed on the employer by the standard nomination procedure, architects often attempt to 'name' sub-contractors under JCT 80 without actually 'nominating' them under clause 35. This practice is considered in detail under the heading 'Naming under other JCT forms', but in the context of discussing nomination, it is important to point out that clause 35 of JCT 80 makes naming without the consequences of nomination very difficult to achieve unless major amendments have been made to the main contract.

This is because of the way in which clause 35.1 of JCT 80 defines 'a Nominated Sub-Contractor'. A sub-contractor will be 'Nominated' within the meaning of JCT 80, with all the contractual consequences that entails, where the architect

- in the bill of quantities/specification/schedules of work, *or*
- in an instruction as to the expenditure of a provisional sum, *or*
- in an instruction requiring a variation, *or*
- by agreement between the contractor and the architect on behalf of the employer

has, either by the use of a prime cost sum **or** by naming a sub-contractor, reserved to himself the final selection and approval of that sub-contractor.

The breadth of that definition means that if an architect simply names a sub-contractor in the main contract documents, that sub-contractor will automatically be 'Nominated' within the meaning of JCT 80 unless clause 35 has either been deleted altogether or has been substantially amended. That sub-contractor could therefore demand to be employed on the basis of NSC/ C, and the employer could find that, among other things, he is bound to grant the main contractor an extension of time if the sub-contractor is in delay, which would be the exact opposite of the result the architect set out to achieve by naming rather than nominating.

Naming under IFC 84

IFC 84 does not provide for nomination, but it does allow the employer to select sub-contractors through the process of naming. Sub-contractors may be named either in the main contract documents, or in an instruction to expend a provisional sum.

Key features of naming

The key features of naming are as follows:

- the sub-contractor is selected by the architect and named in either the main contract documents or in an instruction to expend a provisional sum;
- the standard tendering procedure set out in clause 3.3 of IFC 84, which provides for the use of a standard form of tender, NAM/T, must be used;
- the use of the standard named sub-contract, NAM/SC, is also mandatory;
- by an express provision in the main contract, the main contractor will not be liable for any design provided by the named sub-contractor;
- the named sub-contract work is priced by the main contractor at his own rates and prices – and is not the subject of a prime cost sum;
- once appointed, the named sub-contractor is treated in a very similar way to an ordinary domestic sub-contractor, so the architect is not involved in matters such as certifying payments to him or deciding upon his entitlement to extensions of time;
- the main contractor is not entitled to an extension of time for delays caused by named sub-contractors;
- in the event of the determination of the named sub-contractor's employment, an extraordinarily complex set of provisions will apply, which result in a sharing of the risk between the main contractor and employer.

The naming procedure

The procedure for naming a sub-contractor under IFC 84 varies according to whether the sub-contractor is being named in the main contract documents (and has therefore been appointed before the main contractor), or in an instruction to expend a provisional sum (in which case the main contractor will already be on board).

If the sub-contractor is **named in the main contract documents**, the naming procedure is as follows:

1. The architect completes the invitation to tender in Section 1 of NAM/T and sends it to those being invited to tender for the sub-contract work, together with drawings and either a specification, schedule of works or a bill of quantities describing the work to be carried out.
2. The tenderers complete Section 2 of NAM/T giving details of their tender, and return it to the architect.
3. The architect selects the sub-contractor to be named, and countersigns Section 2 of the NAM/T submitted by the selected tenderer.
4. When tenders are invited for the main contract work, tenderers should be provided with a description of the work to be executed by the named sub-contractor, together with the completed Sections 1 and 2 of NAM/T.

5.   Within 21 days of entering into the main contract, the main contractor is required by clause 3.3.1 of IFC 84 to execute the sub-contract articles of agreement in Section 3 of NAM/T, which incorporate by reference the standard sub-contract conditions NAM/SC.

6.   The contractor must notify the architect of the date he entered into a sub-contract with the named person.

A notable omission from this version of the naming procedure is that the main contractor has no express right of reasonable objection to the named sub-contractor's identity. The theory behind this is that the main contractor had all of the information regarding the proposed sub-contractor when he tendered, and therefore he was able to take account of it in his price. If he was not prepared to work with the proposed sub-contractor at all, he had the option of either qualifying his tender or not tendering at all.

Whilst it is true that Section 2 of NAM/T sets out not only the sub-contractor's tender price, but also the period he requires to carry out his work and the special attendances he wishes the main contractor to provide, these details are unlikely to discourage the main contractor from tendering for the main contract, particularly if the named sub-contract work is not a major element of the building.

The fact that there will inevitably be some negotiation on the named sub-contractor's tender once the main contractor has been appointed is reflected in clause 3.3.1 of IFC 84, which lays down rules to be applied if no sub-contract can be entered into as required by Stage 5. In that event, the contractor must immediately inform the architect, and specify what is preventing agreement.

Provided the architect is satisfied that it is the particulars given in the main contract documents which are preventing agreement (for example the sub-contractor's programme or attendance requirements) the architect must issue an instruction which either:

-   changes the particulars so as to remove the problem (for example agrees to provide an item of special attendance to the sub-contractor himself or amends the main contract completion date); *or*
-   omits the work (in which case the employer could employ the proposed sub-contractor direct); *or*
-   omits the work from the contract documents and substitutes a provisional sum (so he would have the option of naming another sub-contractor to carry it out using the other naming procedure described below).

If the architect takes either of the first two options, the instruction is treated as a variation and the main contractor therefore recovers any additional costs he incurs as a result, together with an extension of time and loss and expense if appropriate.

If the sub-contractor is **named in an instruction to expend a provisional sum**, the procedure is somewhat different, although the initial stages are the same:

1. The architect completes the invitation to tender in Section 1 of NAM/T and sends it to those being invited to tender for the sub-contract work, together with drawings and either a specification, schedule of works or a bill of quantities describing the work to be carried out.
2. The tenderers complete Section 2 of NAM/T giving details of their tender, and return it to the architect.
3. The architect selects the sub-contractor to be named, and countersigns Section 2 of the NAM/T submitted by the selected tenderer.
4. The architect issues to the main contractor an instruction to expend a provisional sum by appointing the selected tenderer and sends with that instruction a description of the work to be carried out together with Sections 1 and 2 of NAM/T and the documents on which the sub-contractor based his tender.
5. The main contractor has a right of reasonable objection to the subcontractor, which must be exercised within 14 days of the issue of the instruction.
6. If the main contractor does not make a reasonable objection, he must execute Section 3 of NAM/T within 14 days of receiving the instruction.

In this version of the naming procedure, IFC 84 does not provide for what happens if the sub-contractor and the main contractor are unable to agree, presumably relying on the right of reasonable objection to solve any problems. Although it would be open to the architect to issue a variation instruction under the main contract to, for example, entitle the main contractor to payment for providing a particular item of attendance to the subcontractor, he is under no obligation to do so, and therefore the system may simply break down.

Although design by named sub-contactors is not dealt with in any of the standard documents issued by the JCT itself, RIBA and CASEC have published a design agreement, ESA/1, for use between named sub-contractors and employers. This is considered in detail in Chapter 4.

Advantages and disadvantages of naming

As with nomination, the benefits and risks of naming vary according to the perspective from which they are being assessed.

The **advantages** of naming for **the employer** are:

(i) the ability to control who carries out key elements of the work;
(ii) the ability to appoint sub-contractors at an early stage, even before the appointment of the main contractor, so that they can contribute to the design process;

    (iii) a direct link with the sub-contractor in respect of design through ESA/1;

    (iv) no responsibility for the performance of the named sub-contractor unless his default or insolvency causes the determination of his employment.

The **disadvantages** of naming for **the employer** are:

    (i)   the main contractor will have no liability for any design work carried out by the named sub-contractor, therefore if the employer cannot implement his direct right against the sub-contractor under ESA/1 because of, for example, his insolvency, he will be without a remedy;

    (ii) if the sub-contractor's employment is determined, the employer will bear the majority of the extra costs incurred as a result (full details of how the costs are allocated are given in Chapter 11).

The **advantages** of naming for **the main contractor** are:

    (i)   he will not be liable for any design done by the named sub-contractor;

    (ii) he bears only limited risks in the event of the determination of the named sub-contractor's employment.

The **disadvantages** of naming for **the main contractor** are:

    (i)   he loses the power to select who carries out the relevant portion of the work;

    (ii) he will have no right to impose his own terms and conditions on the named sub-contractor;

    (iii) any involvement of the sub-contractor in the design process may encourage him to communicate directly with the professional team on other matters, which can create management problems for the main contractor.

The **advantages** of naming for **the sub-contractor** are:

    (i)   the mandatory use of a standard tendering procedure and tender documents which removes any risk of Dutch auctioning;

    (ii) the mandatory use of standard sub-contract conditions NAM/SC;

    (iii) acknowledgement of and, in theory at least, payment for the sub-contractor's input into the design process.

There are no real **disadvantages** for **the sub-contractor** in naming compared to being employed as a domestic sub-contractor, which is the only other method of procurement in IFC 84, but there are disadvantages compared to his position when he is a nominated sub-contractor under JCT 80:

    (i)   payment to named sub-contractors is not related to architect's certificates, but is made on a similar basis to that under DOM/1, the standard form of domestic sub-contract;

(ii)   the architect is not involved in decisions on the sub-contractor's entitlement to extensions of time or in the valuation of loss and expense.

Naming under JCT 81

The only JCT standard form other than IFC 84 to provide for the naming of sub-contractors is the design and build form, JCT 81. The relevant clauses were introduced into that contract as part of the 'Supplementary Provisions' added to JCT 81 in 1988 at the instigation of the British Property Federation.

S4, the clause which permits the naming of sub-contractors in the employer's requirements, is extremely brief compared to the nomination and naming provisions in JCT 80 and IFC 84 respectively. It deals with only two aspects of naming: action to be taken if the contractor is unable to enter into a sub-contract with the named person, and the consequences of the determination of the named sub-contractor's employment.

If the contractor is unable to enter into a sub-contract with the named person, he must immediately inform the employer of why he is unable to do so and, provided there is a bona fide reason why no agreement can be reached, the employer is required to do one of three things:

- issue an instruction requiring a change (the term used in JCT 81 for a variation) removing the reason for the inability; *or*
- omit the named sub-contract work and then reinstate it as ordinary work, so that the main contractor can select his own sub-contractor to carry it out; *or*
- omit the named sub-contract work altogether and employ someone direct to carry it out.

In drafting these rules, the JCT appears to have adopted a policy of not giving the employer the right to name another person to carry out the work, or to name anyone after the main contractor himself has tendered (hence the requirement to name in the original employer's requirements). This will ensure that the main contractor is aware of the identity of any named sub-contractors when compiling his tender so that he can price the risk of their non-performance, for which he is fully responsible unless the default is so serious that their employment is determined.

The supplementary provision S4 does not specify any form of sub-contract on which the named sub-contractor must be employed, although the guidance issued by the Tribunal when the Supplementary Provisions were published did refer to DOM/2 as being an appropriate form of sub-contract.

If the main contractor wishes to determine the employment of a named sub-contractor, he must obtain the employer's consent (which must not be unreasonably withheld or delayed). The main contractor then makes his own arrangements for the sub-contract works to be completed, and is paid for doing so as if it were a change in the employer's requirements. He thus recovers any extra costs arising from that completion, and will be entitled to an extension of time and loss and expense if appropriate.

The advantages and disadvantages of naming under JCT 81 for each of the parties involved are now listed.

The **advantages** for **the employer** of naming under JCT 81 are:

(i)   the power to select sub-contractors through naming them in the employer's requirements;
(ii) no procedural requirements such as the use of a standard form of tender;
(iii) responsibility for default by the named sub-contractor only if his employment is determined.

The **disadvantage** of naming under JCT 81 for **the employer** is:

(i)   full financial responsibility for any additional costs incurred by the contractor as a result of the determination of the named sub-contractor's employment.

The **advantages** for **the main contractor** of naming under JCT 81 are:

(i)   the right to recover additional costs arising out of the determination of the named sub-contractor's employment (which would occur automatically on the named sub-contractor's insolvency);
(ii) freedom to use whatever terms and conditions of sub-contract he can negotiate with the named sub-contractor (although it should be borne in mind that a uniquely named sub-contractor enjoys a comparatively strong negotiating position).

The **disadvantages** of naming under JCT 81 for **the main contractor** are:

(i)   he loses the power to select his own sub-contractors;
(ii) he nevertheless remains responsible for their performance unless their employment is determined.

The **advantages** of naming under JCT 81 for **the sub-contractor** are:

(i)   he is selected by the employer rather than the main contractor, therefore should be protected from practices such as dutch auctioning;
(ii) although he is not entitled to demand to be employed on a standard form of sub-contract, he will be in a strong negotiating position when agreeing terms of sub-contract with the main contractor.

The **disadvantages** of naming under JCT 81 for **the sub-contractor** are:

(i)   he has no contractual right to insist upon the use of a standard form of sub-contract;

(ii) once selected, he has no contractual right to appeal to the employer over matters such as payment and extension of time.

As will be apparent from the above, naming under JCT 81 is employer selection with what the JCT appears to consider the minimum of contractual regulation, and as such may be the model for naming provisions in other standard forms in the future.

Naming where no main contract provisions exist

The ability to choose a sub-contractor without assuming any responsibility for his performance or solvency is an attractive proposition for most employers, and therefore sub-contractors are frequently named in the documents describing the work to be carried out by the main contractor, regardless of any provisions on sub-contracting in the main contract.

As already mentioned, this is a dangerous practice under JCT 80, as any sub-contractor chosen by the architect will be treated as a 'nominated sub-contractor' under clause 35 unless, of course, that clause has either been substantially amended or deleted altogether.

Similarly in IFC 84 or JCT 81, a sub-contractor named in the main contract documents will be subject to the rules on named sub-contractors set out in those contracts provided, in the case of JCT 81, that the Supplementary Provisions are stated to apply in the Appendix to the main contract.

Naming sub-contractors under forms such as the Agreement for Minor Building Works may be more successful in the sense that there are no express provisions which prohibit it, or which stipulate consequences of doing so which are unattractive to employers.

However, there are more serious issues to be considered in relation to naming under non-standard provisions than simply how difficult it is to achieve from a drafting point of view.

Is it, for example, fair to expect a main contractor to take full responsibility for a sub-contractor whom he did not choose, and whose identity he may not even have known when he tendered for the work? Is it right that the sub-contractor should be deprived of all the normal benefits of employer selection, including a mandatory standard form of sub-contract? And how is any design work done by the sub-contractor to be handled? Contracts such as JCT 80, IFC 84 and the Minor Works Form are all contracts for work and materials only, where a strong argument can be put forward that the contractor has no responsibility for design done by his sub-contractors, even if there is no express exclusion of design responsibility in the main contract on which the main contractor can rely.

What is needed is a system of naming in all of the JCT standard forms, particularly in JCT 80, which represents a fair balance of risk between the main contractor and employer, and which gives the sub-contractor both protection against unfair tendering practices and the benefit of a standard form of sub-contract. Nomination has obviously not answered that need, as many employers have voted with their feet on clause 35 of JCT 80. It is to be hoped that the JCT can eventually provide the type of deal on naming which the industry so obviously, and urgently, requires.

List of three – clause 19.3 of JCT 80

Clause 19.3 of JCT 80 provides a halfway house between nomination and domestic sub-contracting by allowing the employer to include in the main contract documents a list of people from whom the main contractor must select a sub-contractor to carry out a particular portion of the works.

The key features of the list of three procedure are:

- a list of names is included in the contract documents from whom the main contractor must choose the sub-contractor;
- the work is priced by the main contractor in the ordinary way;
- there are no standard tender forms or standard form of sub-contract which must be used;
- once the sub-contractor has been appointed he is an ordinary, domestic sub-contractor, and therefore the architect does not participate in the administration of the sub-contract in any way whatsoever;
- the main contractor is fully responsible for the performance of the sub-contractor he selects, and bears all of the risks if his employment is determined.

The list of three procedure

The list of three procedure is both brief and straightforward. It works as follows:

1. The architect states in the main contract documents that certain work must be carried out by a sub-contractor selected from a list of at least three names included in those documents.
2. The main contractor seeks tenders for that work from the persons named on the list when and how he chooses (although he is not contractually obliged to seek a tender from every sub-contractor on the list).

3. At any time before a sub-contract is entered into, either the employer (or the architect on his behalf) or the contractor, with the consent of the other, may add names to the list.
4. If less than three of the people named on the list are able and willing to carry out the sub-contract work, the employer and contractor may agree names to be added to make the list up to three again, or the main contractor can let the work as an ordinary domestic sub-contract to anyone he chooses.

This system gives the employer no more than an element of control over who carries out the sub-contract work, and main contractors often circumvent the list altogether by persuading the architect to accept the addition to the list of their own favoured sub-contractor.

The clause 19.3 procedure is unpopular with sub-contractors because it gives them no protection during the tendering process, and does not guarantee the use of a standard form of sub-contract. It also fails to acknowledge contractually any input the sub-contractor may have had into the design process.

Selecting works contractors under the Management Contract

It is consistent with the theme of co-operation which runs throughout the JCT Management Contract that it provides for works contractors to be selected by agreement between the employer and the management contractor and for that agreement to be confirmed in an architect's instruction.

The key features of the selection process stipulated by the management contract are:

- works contractors are selected by agreement in writing between the management contractor and employer;
- the standard works contract documentation issued by the JCT must be used unless the employer (or the architect on his behalf) and the management contractor otherwise agree;
- the standard documentation includes a form of tender and agreement, Works Contract/1, which must be completed at some stage, although the management contractor is not bound to use it to obtain tenders;
- standard works contract conditions must be used unless the employer and management contractor have agreed otherwise;
- the tender process is administered by the management contractor;
- decisions such as the amounts to be paid to works contractors and when practical completion of their work is achieved are made by the architect;
- the management contractor awards extensions of time to works contractors. (Although the architect has the right to register a dissent if he

does not agree with the decision. Such a dissent will not affect the works contractor's entitlement, but may be relevant in deciding whether the management contractor has properly discharged his role under the management contract.);

- the management contractor is prima facie responsible for defaults of his works contractors, but is relieved of the consequences of those defaults by a complex set of relief provisions.

The procedure for selecting works contractors

The JCT Management Contract envisages that contracts for the various packages of work will be placed progressively during the project, to allow maximum flexibility and an early start on site. Stages 3 to 8 of the process now described will therefore be carried out both during the pre-construction period, and after work on site has commenced.

1. During the pre-construction period (i.e. before any work is carried out on site), the management contractor advises the professional team on the breakdown of the project into suitable packages of work.
2. At that stage, he is also required to prepare tender lists and to investigate and report on the capabilities and financial standing of the tenderers. (This often takes the form of extensive pre-tender interviews.)
3. The management contractor invites tenders for the various works contracts. He is not required to use a standard document to do so, but he may use the standard invitation in Section 1 of Works Contract/1 if he so wishes.
4. When he invites tenders, the management contractor must state whether the successful tenderer will be required to enter into the standard employer/works contractor agreement, Works Contract/3, which creates direct liability to the client in respect of design.
5. If Works Contract/1 has been used to invite tenders, the tenderers must complete Section 2 of that document, giving details of their price and other matters such as daywork percentages, additional attendances required and proposals for the period required to execute the work. If the standard invitation was not used, the tenderers simply tender in the manner specified in and giving the information required by the main contractor's invitation.
6. The management contractor agrees with the architect who should be appointed as works contractors, and that agreement is confirmed by the issue of an instruction by the architect to the management contractor.
7. On receipt of that instruction, the management contractor arranges for the completion of Sections 1 and 2 of Works Contract/1 if they have not already been completed, and checks there are no discrepan-

cies between the two sections on matters such as programme and attendances.

8. The management contractor and the works contractor execute the articles of agreement in Section 3 of Works Contract/1, and if required to do so, the works contractor enters into Works Contract/3 with the employer.

The need to ensure that there are no discrepancies between Parts 1 and 2 of Works Contract/1 mentioned in Stage 7 arises because both documents are annexed to and become part of the contract between the parties set out in Section 3 of Works Contract/1. All three sections of Works Contract/1 are prefaced by a warning that there must be no discrepancies, and the management contractor must therefore reach agreement with the works contractor on matters such as the period required to carry out the work both off site and on site and attendances before Section 3 is executed.

Unlike other JCT forms, the Management Contract does not provide for the architect or employer to participate in this process, and perhaps to resolve a deadlock by instructing the management contractor to provide certain items.

Attendances can be a particular problem on management contracts, as the management contractor himself will be carrying out no work on site. The management contractor will normally be required to provide certain items of general attendance which are listed in the Fifth Schedule to the Management Contract, but that will be a matter for agreement between the management contractor and the employer before the beginning of the construction period. No minimum requirements are stipulated in the standard form, although a model checklist including most general attendance items is set out in a JCT guidance note.

Works Contract/1 also contains a printed list of attendance items in Section 1, but it will be open to the management contractor to amend the printed list so that it corresponds with the Fifth Schedule to his own contract. The attendances offered to works contractors may therefore be minimal.

The works contractor is entitled to ask for other attendances to be provided, but these will not be free of charge: clause 3.11.2 of Works Contract/2 requires him to pay either the sum agreed or a fair and reasonable price for such extra attendances.

As the Management Contract contains a single set of rules for appointing works contractors, an evaluation of its advantages and disadvantages is unnecessary.

Domestic sub-contractors

All of the JCT standard forms, with the exception of the Management Contract, provide for the appointment of ordinary domestic sub-contractors. They also contain a provision that the main contractor may not sub-let any portion of the main contract works without the written consent of the architect, which must not be unreasonably withheld.

The forms do not specify whether consent must be given simply to the fact of the sub-letting, i.e. the carrying out of work by someone other than the main contractor, or to the identity of the proposed sub-contractor. It is, however, generally accepted that the employer's consent must be obtained to the identity of the proposed sub-contractors, although in practice the right to object is rarely exercised.

The work to be carried out by domestic sub-contractors is priced and paid for by the employer at the main contractor's own rates and prices, and the amount paid to the sub-contractors is dictated solely by the terms of sub-contract negotiated between the parties to it.

The main contractor is fully responsible for the quality of work provided by domestic sub-contractors, and for any delays which they might cause. If they become insolvent, thereby causing delay and extra costs, that is also entirely at the main contractor's risk.

The only potential exception to this rule of total responsibility for any default by domestic sub-contractors is in relation to design. Contracts such as JCT 80 and IFC 84 do not require the contractor to carry out any design work (unless the contractor's designed portion supplement is used or non-standard amendments are made). It can therefore be argued that the main contractor is not responsible for the quality of any design done by his domestic sub-contractors, as he does not owe any obligations to the employer under the main contract in that respect. This is a complex issue on which there are no clear rules, and it is discussed in more detail in Chapter 4.

There is no mandatory form of domestic sub-contract, although all the JCT forms of main contract require that two provisions must be included in all domestic sub-contracts, namely:

- that the employment of the domestic sub-contractor will determine automatically on the determination of the main contractor's employment under the main contract; *and*
- where the value of goods and materials has been included in a certificate under the main contract and the amount due under that certificate has been discharged by the employer, the sub-contractor undertakes not to deny that those materials and goods have become the property of the employer. (The reasons for this requirement are discussed in Chapter 5.)

BEC, FASS, CASEC and FBSC publish standard forms of domestic sub-contract for use with JCT 80, IFC 84 and JCT 81, but their use is entirely optional. Most large main contractors have their own standard terms and conditions of sub-contract which they try to use as often as possible. Some of these forms are based on documents such as DOM/1, but with extensive amendments to clauses dealing with matters such as payment and set off; others are drafted on a completely unique basis.

The bodies representing sub-contractors on the JCT have periodically pressed for the adoption of the standard forms of domestic sub-contract by the JCT, but this has so far been strongly resisted by representatives of both main contractors and private sector clients.

Sub-sub-contractors

Sub-sub-contractors are those persons to whom the sub-contractor in his turn lets part of the sub-contract work. The sub-contractor is in a similar position to the main contractor in that such sub-letting is permitted under the standard forms of sub-contract only with the consent of the contractor, which must not be unreasonably withheld. In the case of sub-letting under the nominated sub-contract NSC/C, the consent of the architect is also required.

There are no restrictions on the way in which sub-contractors seek tenders from their sub-sub-contractors, and no requirements as to the terms and conditions which must be used.

The sub-contractor will be fully responsible for the proper and timely execution of any work by his sub-sub-contractors, although the same question mark over liability for design arises here as arises in relation to a main contractor's liability for design done by domestic sub-contractors.

It is sometimes claimed that sub-contractors pursue a policy of 'order above and chaos below' with their sub-contractors – in other words they expect standard forms of sub-contract and fair tendering procedures between themselves and the main contractor, but also want complete freedom to obtain the maximum commercial advantage from their sub-sub-contractors. How true this is varies from sub-contractor to sub-contractor, but it is inevitably the case that the lower down the contractual chain a company is, the harder it has to fight for fair treatment.

The only exception to this rule is materials suppliers, who do not carry out any work on site and who seem to be able to secure favourable conditions regardless of how harshly the sub-contractors and sub-sub-contractors are being treated. Their position is discussed further in Chapter 5.

4 Design by Sub-Contractors

The days when designers designed and builders built are long gone. Creating a modern building involves a range of specialist design skills which a single architect could not possibly master, even with the help of other members of the professional team such as the services and structural engineers. As a result, many specialist sub-contractors not only provide work and materials – they also design an element of the building.

This is particularly true of product based sub-contractors like lift contractors and fire alarm specialists, the majority of whose work is done off site in their own factories with on-site installation representing only a fraction of their sub-contract value.

Another fertile area for sub-contractor design is mechanical and electrical services, which can account for up to half the value of a sophisticated building. It is almost unheard of for such services to be fully designed by the professional team, and frequently the sub-contractor is supplied with no more than a performance specification. The traditional method of dealing with such a high level of design input from sub-contractors was nomination, but with the decline in its popularity new and often convoluted ways of dealing with design input from sub-contractors have evolved.

The purpose of this chapter is to describe the different ways in which sub-contractors become involved in the design process, to examine the standard forms of contract which cater for design by sub-contractors, and to look at other ways in which they may become legally liable for design. Several design-related issues such as copyright in design drawings and professional indemnity insurance are dealt with at the end of the chapter.

When do sub-contractors design?

Few sub-contractors work to precise, detailed drawings supplied by the professional team which tell them exactly how they should carry out their work. But this does not, in itself, mean that all sub-contractors are liable for design. The distinction between design and workmanship is a difficult one to draw, and it can only sensibly be attempted by reference to practical examples.

For instance, when a sub-contractor chooses what size of nail to use to fix a bracket to a wall it is most unlikely that he is designing; when he determines the precise positioning of the brackets to fix a pipe to a wall he may be designing; if he is selecting the size of air handling unit which should be used to maintain a specified temperature he is definitely designing.

Ultimately, what constitutes design will be a question of fact in every case, but if the sub-contractor is responsible for preparing plans, specifications or drawings (other than as-installed drawings, which simply *record* what he has installed rather than *dictate* what he is to install) or if he is selecting major items of plant or equipment, there is a strong inference that he is designing.

Some of the most common ways in which sub-contractors become involved in design are as follows:

- when the sub-contract works are required to meet a performance specification but no detailed drawings or instructions as to how that specification should be met have been supplied by the professional team;
- when the professional team is appointed to carry out what their standard conditions of engagement refer to as abridged duties, where they produce only outline and schematic drawings, so the sub-contractor must produce his own detailed drawings;
- where the main contractor is appointed on a design and build basis and all sub-contractors are required to design their own sub-contract works;
- when the sub-contractor is providing a specialist product such as a lift or curtain walling system which he has designed.

Contracts providing for design

The vast majority of the standard forms issued by the Joint Contracts Tribunal are drafted on the basis that the main contractor will not be carrying out any design. There are only three exceptions to this rule: the With Contractor's Design Form 1981, the Contractor's Designed Portion Supplement for use with JCT 80 and the performance specified work provisions in part 5 of JCT 80.

There are, however, many more JCT agreements which provide for design by sub-contractors, a fact which reflects the importance of their design input on most projects. Some of these agreements are with the employer rather than the main contractor, particularly where it is the employer who has selected the sub-contractor as in nomination under JCT 80 or naming under IFC 84.

The terms relating to liability for design in each of these contracts are now considered.

JCT with Contractor's Design Form 1981

JCT 81 is drafted in such a way that the main contractor may be responsible for designing the whole of the works, or for completing a design partially prepared by an independent professional designer appointed by the employer. The degree of design which the contractor is required to undertake will be apparent from the Employer's Requirements, which can be anything from a simple description of the accommodation required to a full set of schematic drawings.

Skill and care

Clause 2.5.1 of JCT 81 sets out the contractor's design warranty. In essence, it provides that the contractor's liability for design is the same as would be imposed upon an architect or other appropriate professional designer, which means that the contractor must exercise reasonable skill and care in the preparation of the design.

Fitness for purpose

A professional designer would not usually be responsible if the completed building was not fit for its intended purpose, provided he had discharged his duty of skill and care. However, the courts have held that if a contractor both designs and constructs, he will be under an implied obligation to produce a building which is reasonably fit for its purpose (see, for example, the case of *Viking Grain Storage* v. *T. H. White Installations Ltd* (1985) 33 BLR 103).

A fitness for purpose obligation is very much more difficult to discharge than a duty to exercise reasonable skill and care, because it involves guaranteeing results. If the contractor is required to produce a building which is fit for its purpose, all the employer will need to prove in order to recover damages for breach of contract is that the building is defective in some way so that it is unfit for its purpose, regardless of why that may be the case. He will not need to prove that the contractor was in default in any way whatsoever.

Clause 2.5.1 of JCT 81 therefore attempts to exclude the implied term as to fitness for purpose and to put a contractor working under that contract in the same position as a professional designer. It does not expressly refer to fitness for purpose, but states that the design provided by the contractor should be treated as though it were supplied under a separate contract with the employer from that for the construction of the building.

It is sometimes argued that, because clause 2.5.1 purports to alter the common law position to the detriment of the employer, it could be attacked under the Unfair Contract Terms Act 1977. Despite its name, that Act does not outlaw all unfair contract terms, but section 3 provides that where one party contracts on another's written standard terms of business, liability for breach of contract can only be excluded or restricted to the extent that the relevant term satisfies the requirement of reasonableness.

An employer seeking to challenge clause 2.5.1 will therefore have two hurdles to overcome, as he will have to prove that:

- the JCT Design and Build form is the contractor's written standard terms of business; *and*
- clause 2.5.1 does not satisfy the requirement of reasonableness.

Both of these will be difficult. Although there is no direct authority on the point, most commentators believe that a JCT standard form should not be treated as either party's standard terms of business, because it is negotiated by bodies representing all sides of the industry, including clients and contractors.

But even if a court were to take a different view, a term excluding fitness for purpose may satisfy the requirement of reasonableness. This is because one of the key tests of whether an exclusion is reasonable is whether the person seeking to exclude liability could have insured against it. Insurance against faulty design is generally only available to contractors for a breach of the duty to exercise reasonable skill and care, not for a failure to provide a building which is fit for its purpose. Other factors, such as the relatively equal bargaining power of the parties and the frequency with which such a term is accepted in the building trade will also militate against a finding that the clause is unreasonable.

Financial limitation on liability

JCT 81 also provides in clause 2.5.3 for a financial limitation on the contractor's liability for defective design. However, this clause will only operate where a sum to which the contractor's liability will be limited is stated in the appendix, and in practice it is very rarely used.

In any event its ambit is severely restricted, as it will not operate to the extent that the works are concerned with the provision of a dwelling because this would be contrary to the Defective Premises Act 1972. In addition, it will not affect the contractor's liability for liquidated damages for delay, only his liability for other forms of consequential loss such as loss of profit or loss of use.

Contractor's Designed Portion Supplement

The Design and Build Form was not the only new contract produced by the JCT in 1981: in the same year a supplement for use with JCT 80 was published to cater for the design by the contractor of a part of the works.

The Contractor's Designed Portion Supplement 1981 (the CDP Supplement) contains provisions which are very similar to those in JCT 81. For example, the contractor's obligations in respect of the design of the part of the works covered by the CDP Supplement are exactly the same as those imposed on the contractor by clause 2.5.1 of JCT 81, so the comments made above regarding reasonable skill and care, fitness for purpose and financial limitation of liability are all equally relevant here.

In addition, the architect has an express obligation to issue directions to integrate the contractor's design with the design of the rest of the works. This is helpful in that it makes it clear that the architect retains overall responsibility for the co-ordination of design, but the use of the word 'direction' rather than 'instruction' implies that the contractor may be unable to recover any additional costs he incurs in complying with such directions. The contractor may be able to argue that he is entitled to extra money if he can show that the 'direction' amounts to a change in the Employer's Requirements in respect of that part of the work which is subject to the CDP Supplement, but it is not clear whether a court would accept such a proposition.

In order to ensure that the contractor retains control over the design of the part of the works which is subject to the CDP Supplement, clause 2.7 provides that if he is of the opinion that any architect's instruction or direction will 'injuriously affect' the efficacy of his design, he can give the architect notice to that effect, and he need not comply with the instruction unless it is confirmed by the architect. In that event, the contractor would presumably be relieved of liability for any problem with the design of which he had warned in his notice.

Clause 2.8 of the Supplement obliges the architect to give notice to the contractor of anything which appears to him to constitute a defect in the contractor's design. The contractor is not bound to take any action on receipt of such a notice, but he would obviously be foolish to ignore it unless he was certain that the architect was wrong.

Although clause 2.8 is drafted in such a way as to make it clear that the architect **must** warn the contractor of anything which, in his opinion, constitutes defective design, it will not allow the contractor to escape liability for any defects which the architect does not point out. This might lead some architects (and even contractors) to query the usefulness of this clause, but it is suggested that it serves an important practical purpose by making it clear that the architect cannot wilfully ignore any errors in the contractor's design.

The CDP Supplement is drafted for use only with the With Quantities versions of JCT 80, so further modifications will be necessary if the employer wishes to use it on a project where there is no firm bill of quantities.

Use of the CDP Supplement in relation to sub-contract work

As the popularity of nomination has declined, so the use of the CDP Supplement has increased. A common way to deal with sub-contractor design is to delete clause 35 of JCT 80 which provides for nomination, to name a sub-contractor in the main contract documents, and to make the sub-contract works the subject of the CDP Supplement. In this way, the employer makes the main contractor fully responsible for the sub-contractor's design, and avoids any liability for the sub-contractor's late completion or insolvency, both of which are at his risk where the sub-contractor is nominated.

There are, however, several drawbacks to this scheme, particularly for the main contractor. One of the most significant is that there is no CDP Supplement for use with DOM/1, therefore the main contractor will have to draft his own amendments to the sub-contract to step down the design and other obligations imposed upon him by the CDP Supplement at main contract level.

A popular, but inadvisable, tactic in these circumstances is to employ the sub-contractor on the basis of DOM/2 – this will impose obligations on the sub-contractor regarding design, but will also import other terms into the sub-contract which step down terms of the JCT Design and Build form (for which DOM/2 is designed) which are not relevant to JCT 80 as modified by the CDP Supplement. Although it is more time-consuming to step down each clause of the CDP Supplement into DOM/1, this is by far the best approach.

Performance specified work – part 5 of JCT 80

Part 5 of JCT 80 was added to the contract by Amendment 12, which was published in July 1993. It provides for the design by the contractor of certain items of work which will be identified in the Appendix, although the term 'design' is carefully avoided in the contract itself.

The description of the work which is subject to clause 42 given to the contractor at tender stage must be sufficiently detailed to enable him to price the work. If it is not, a provisional sum must be included in the contract bills together with information regarding the performance required from the work, its location and enough information for the contractor to have allowed for programming it and pricing preliminaries related to it.

Unlike work covered by the CDP Supplement, the contractor need not submit his proposals for carrying out the performance specified work at tender stage, although he must do so by a date which will either be included in the contract bills or in an architect's instruction. If no date is stipulated, the proposals must be submitted a reasonable time before the contractor carries out the work. Delay in providing these proposals, described in clause 42 as 'the Contractor's Statement', will involve the payment of liquidated damages if it causes a delay to the project.

Under clause 42.17 the contractor is obliged to exercise all reasonable skill and care in the provision of performance specified work and there is, for the first time in a JCT form, an express exclusion of any guarantee as to fitness for purpose. In view of the debate which surrounds this issue under the other JCT documents, this is a precedent which could usefully be followed elsewhere.

Another helpful aspect of clause 42 is that it makes it absolutely clear that the architect retains responsibility for the overall co-ordination of the design, as clause 42.14 requires him to issue instructions for the integration of the performance specified work into the rest of the works.

Clause 42 expressly prohibits the supply of performance specified work by nominated sub-contractors, but it will often be supplied by domestic sub-contractors, and it is understood that an amendment to DOM/1 to cater for this was in preparation in 1993/94.

JCT Management Contract 1987

Although the management contractor does not himself carry out any design, he is liable (subject to the relief provisions) for any breach of the works contracts by the works contractors, and as the standard works contract Works Contract/2 includes obligations relating to design, the management contractor will be exposed to claims from the employer in the event of defective design by works contractors. The liabilities imposed on the sub-contractor by Works Contract/2 in respect of design are described later in this chapter.

Sub-contractor's liability for design

It is important to appreciate that, in the event of defective design by a sub-contractor whose sub-contract contains express terms regarding design, such as DOM/2, if the sub-contractor's design is defective, the employer's remedy is against the main contractor, who in turn will claim against his sub-contractor. In such circumstances, unless the employer enters into a separate, non-standard warranty with the sub-contractor in respect of design, he will have no independent right to sue the sub-contractor for breach of contract, and could therefore be left without a remedy if the main contractor has become insolvent or has no financial resources.

This is quite different to the position under other standard forms where the main contractor is not responsible for design. In that event, any contractual obligations regarding design are generally owed directly to the employer, for example under the employer/nominated sub-contractor warranty NSC/W, and do not pass through the main contractor.

The only standard form which gives the employer the choice of suing either the main contractor or the sub-contractor when the sub-contractor has supplied defective design is the JCT Management Contract documentation, which contains a warranty regarding design both in the Works Contract conditions (a breach of which would put the management contractor in breach of the Management Contract) and in the Employer\Works Contractor Agreement Works Contract/3.

Sub-contract DOM/2

DOM/2 is the standard form of sub-contract for use where the main contract is the JCT Design and Build Form 1981.

It is a somewhat unusual sub-contract in that, rather than consisting of a set of articles of agreement which incorporate its own standard conditions, it consists of articles of agreement which incorporate by reference the conditions applicable to DOM/1, subject to a schedule of amendments to make them suitable for use where the main contract is JCT 81. The amendments made to the Design and Build Form in recent years have made it a very different form to JCT 80, and so the length of this schedule of amendments has gradually increased. However, the key distinction between the DOM/1 and DOM/2 conditions remains the addition of a clause relating to design.

The basic obligation in relation to design imposed by clause 5.3.1 of DOM/2 is in exactly the same terms as that in JCT 81, and the sub-contractor is therefore obliged to exercise the degree of skill and care expected of an appropriate professional designer. This may be an architect, but is just as likely to be a specialist designer such as a consulting engineer. As under JCT 81, there should be no obligation as to fitness for purpose, as the sub-contractor is deemed to have supplied his design under a separate contract from that for the construction of the works. (For a full discussion of fitness for purpose, see the earlier discussion of the design liability imposed by JCT 81.)

DOM/2 does not contain a standard clause permitting the sub-contractor to negotiate a limitation on his liability for consequential loss in the event of defective design, but clause 5.3.3 states that where such a limitation has been agreed under the main contract, the sub-contractor will receive the benefit of that arrangement because the main contractor will only be entitled to recover damages from him in respect of defective design to the extent that he is bound to pay those damages to the employer under the main contract.

Employer/nominated sub-contractor agreement NSC/W

Because clause 35.21 of JCT 80 expressly relieves the main contractor of liability for defective design by nominated sub-contractors, a direct contractual link in respect of design between the employer and nominated sub-contractor is clearly essential: this is provided by the employer/nominated sub-contractor agreement NSC/W.

Clause 2.1 of that agreement requires the nominated sub-contractor to exercise all reasonable skill and care in:

- the design of the sub-contract works, insofar as they have been or will be designed by the sub-contractor;
- the selection of materials and goods for the sub-contract works, insofar as they have been or will be selected by the sub-contractor;

- the satisfaction of any performance specification or requirement, insofar as it has been included or referred to in the description of the sub-contract works.

At first sight, this appears to place upon the nominated sub-contractor a very similar liability to that imposed by JCT 81 and DOM/2. However, there are some important differences. For example, although the duty imposed on the sub-contractor is one of reasonable skill and care, there is no express reference to the standard by which this should be judged. Is it that of a reasonably competent professional designer, or that of a specialist sub-contractor supplying design? In most cases they are likely to be the same, but there may be instances where a higher degree of care would be expected of a professional, perhaps because of his professional training or because of his duty to follow guidelines laid down by his professional body.

Secondly, the position regarding fitness for purpose may be different. As explained above, under JCT 81 and DOM/2 the obligation regarding fitness for purpose which would otherwise be implied where the contractor both designs and constructs is probably displaced by a provision that the contractor's design should be treated as though it had been supplied under a separate contract.

Where a nominated sub-contractor is designing, the design is supplied under a separate contract, namely NSC/W, from the work and materials. However, it is sometimes argued that, because fitness for purpose is not expressly excluded, the law will imply a term that, if the sub-contractor is designing the sub-contract works, they must be reasonably fit for their purpose. It is suggested that this is incorrect, because a term can only be implied into a contract if there is no express term which would contradict it. In the case of NSC/W, the parties have chosen the standard of liability for design, namely reasonable skill and care, and have thus excluded fitness for purpose. Sub-contractors should, however, be ready to face arguments to the contrary.

In addition to his obligations as to the quality of his design, clause 3.2 of the warranty requires the sub-contractor to provide design information either in accordance with 'the agreed programme details' or at such time as the architect may reasonably require. If he fails to do so and as a result the main contract works are delayed, the main contractor would be entitled to an extension of time. However, the employer could recover damages in respect of the delay from the nominated sub-contractor on the grounds of a breach of NSC/W, although whether those damages would be calculated in accordance with the rate of liquidated damages specified in the main contract or the employer would have to prove his actual loss is open to question (for a full discussion on this point see Chapter 8 on Extensions of Time).

NSC/W emphasises the importance of the link with the employer in respect of design by allowing the architect to instruct the nominated sub-contractor to carry out design work before a formal nomination instruction has been issued, which may be before the main contractor has even been appointed. If for any reason the nomination does not proceed, the warranty also provides for the sub-contractor to be paid any expenses he reasonably incurs in complying with those instructions.

In addition to its provisions regarding design, NSC/W contains provisions on other matters such as direct payment and early final payment which are of benefit to the sub-contractor.

Employer/Specialist Agreement ESA/1

Like JCT 80, IFC 84 (clause 3.3.7) expressly excludes the main contractor's liability for any design work carried out by named sub-contractors, so a direct contractual link between the employer and the named sub-contractor will be essential if the sub-contractor is doing any design. In the case of named sub-contractors, this link will be provided by the employer/specialist agreement ESA/1.

Although ESA/1 is published by the RIBA and CASEC rather than the JCT, the named sub-contractor's obligations regarding design are expressed in almost identical terms to those in NSC/W, therefore the comments already made regarding clause 2.1 of NSC/W are equally relevant here.

Agreement ESA/1 is also similar to NSC/W in that it provides for design to be carried out and goods and materials to be ordered in advance of the sub-contractor's appointment under NAM/SC. But unlike NSC/W, it is drafted in such a way that it can be used as a pure design agreement, as it gives the employer the option of inviting the prospective named sub-contractor to provide only an 'approximate estimate' for carrying out the named sub-contract work when he enters into ESA/1. This can then be developed into a firm tender price (which could be judged in competition with other tenders) when the design of the named sub-contract works is sufficiently progressed.

ESA/1 obliges the named sub-contractor to supply design information in accordance with certain time requirements, which will either be stated in the agreement itself, and/or in the sub-contract or in the tender documents. If a failure to comply with the time requirements results in a delay in either the completion of the main contract works or, if the sub-contractor has been appointed at a very early stage, in the preparation of the main contract documents, the employer will be entitled to recover damages.

If the main contractor has not yet been appointed, it is difficult to see how such damages could be calculated in accordance with the rate of liquidated damages the employer intends to impose on the main contractor in the event of delay, so he is likely to be forced to quantify and prove his loss in those circumstances.

Works Contract/2 and Works Contract/3

As explained above, under the JCT Management Contract documentation, there are two ways in which an employer could recover damages in respect of defective design carried out by a works contractor:

- he could bring an action against the management contractor on the grounds of a breach of Works Contract/2 by the works contractor; *or*
- he could bring an action against the works contractor directly for a breach of the Employer/Works Contractor agreement Works Contract/3.

The fact that he has a choice is very much to the employer's advantage. For example, if for some reason his rights against the management contractor cannot be exercised, perhaps because the management contractor is insolvent, the employer simply brings an action directly against the works contractor under Works Contract/3.

If, however, it is the works contractor who is insolvent, the employer's right to sue the management contractor for a breach of Works Contract/2 will be less valuable, because the management contractor will be protected by the relief provisions in clause 3.21 of the Management Contract, which permit recovery from the management contractor only to the extent that he has succeeded in obtaining damages from the works contractor. The real advantage to the employer of being able to recover from the management contractor rather than under Works Contract/3 will arise where it is not clear which of several works contractors is responsible for the defective design, as in those circumstances the employer can simply claim damages from the management contractor and leave him to sort out the problem with the various works contractors.

Employers attracted by this option should, however, remember that if legal proceedings against the works contractors are necessary to ascertain who is responsible, the legal costs will be met by the employer, as the management contractor is entitled under clause 3.21 of the Management Contract to be paid the costs of enforcing the terms of the works contracts.

Regardless of whether the action is based upon a breach of Works Contract/2 or Works Contract/3, the duty imposed upon the works contractor in respect of design is the same. The relevant clauses are drafted in identical terms to the design warranty in NSC/W, so the works contractor must exercise all reasonable skill and care in:

- the design of the works insofar as they have been or will be designed by him;
- the selection of materials and goods insofar as he has or will select them;
- the satisfaction of any performance specification or requirement which is included in or referred to in the works contract.

The comments made in respect of the design warranty in NSC/W are therefore also relevant here.

Like NSC/W, Works Contract/3 requires the works contractor to provide
design information to the architect in accordance with any agreed programme or at such time as the architect may reasonably require. As there is
no similar obligation in Works Contract/2, if the employer wishes to bring an
action against the works contractor on the grounds of the time at which the
design information was provided rather than on its content, he will have to do
so under Works Contract/3.

A point of distinction between Works Contract/3 and the other standard
warranties such as NSC/W is that it does not provide for design or ordering of
materials before the execution of the Works Contract. According to the JCT's
Practice Note MC/2 this is a deliberate omission; the Practice Note recommends that if early involvement of a prospective works contractor in design is
required, it should be the subject of a separate agreement.

Liability under contracts which do not provide for design

In many cases where a contractor or sub-contractor is carrying out design,
there will be express terms in either the main contract, the sub-contract or in
an employer/sub-contractor warranty which deal with that design. But this
will not invariably be the case.

For example, a domestic sub-contractor under JCT 80 may be supplied with
a performance specification but no further details as to how he is to carry out
his work so that it will meet the required standard. Neither JCT 80 nor DOM/1
includes any express terms relating to design, so it is necessary to consider
whether the general law will imply any terms regarding the provision of any
design which is necessary to allow the sub-contractor to comply with the
sub-contract documents.

This is a difficult question to answer, and one which has not yet been
seriously addressed by the courts. The contracting side would argue that
both JCT 80 and DOM/1 are contracts for the supply of work and materials
and not for design, and therefore that implied terms, for example as to the
exercise of reasonable skill and care, or as to fitness for purpose, would be
inconsistent with the expressed intentions of the parties. However, JCT 80
only excludes the main contractor's liability for design in respect of that
which is carried out by nominated sub-contractors, and DOM/1 does not
contain any exclusion of design liability at all, so this may be a difficult
argument to sustain.

The employers, on the other hand, are likely to argue that the basic
obligation imposed on both the main contractor and the sub-contractor is
to carry out and complete the works in accordance with the contract
documents, and if either fails to do so they will be in breach of contract. A

failure to comply with a performance specification would be such a failure, and so even if that failure was due to defective design, damages would be recoverable for breach of contract. It could also be argued that the Supply of Goods and Services Act 1982 imposes an obligation as to fitness for purpose wherever the buyer relies on the seller's skill and judgement, and that such reliance occurs whenever the sub-contractor is required to design.

In view of the uncertainty which surrounds this issue, both main contractors and domestic sub-contractors may be well advised to take refuge in the extensions of time and loss and expense provisions of JCT 80 and DOM/1, which provide that both contractor and sub-contractor are entitled to an extension and loss and expense if the architect does not supply them in due time with 'all **necessary** instructions, drawing, details or levels'. The contractor or sub-contractor could therefore request the further design information from the architect which is necessary to meet the performance specification, and if it is not forthcoming, they would probably be fully protected against the consequences of any delay.

It should, however, be borne in mind that this strategy will not endear the contracting side to either the employer or his design team, particularly if it is applied to a part of the works which is traditionally designed by the contractor, such as roof trusses or precast concrete floors.

Duty to warn

Even where a sub-contractor is not designing any part of his work, he is at risk of being caught up in an action for defective design because he may be under a duty to warn the contractor or the professional team of defects in their design. The question of whether such a legal duty exists has vexed the courts for many years, and has not yet been fully resolved.

In the case of *Equitable Debenture Assets Corporation* v. *William Moss Group Ltd* (1984) CILL 74 it was held that a main contractor was under an obligation to inform the architect of his opinion that part of the architect's design would not work, even though there were no express obligations regarding design in the building contract. The judge found that such a duty arose both through an implied term in the contract and in tort.

Although Judge Newey, who heard the *EDAC* case, followed his own decision later the same year in *Victoria University of Manchester* v. *Hugh Wilson and Lewis Womersley* (1985) Con LR 43, later cases have tended to doubt the existence of a duty to warn. For example, in *University Court of Glasgow* v. *Whitfield* (1988) 42 BLR 66 Judge Bowsher reviewed the authorities on the existence of a duty to warn and concluded that, where there was a detailed contract between the parties (in this case a Scottish version of a JCT standard form), there was no room for the implication of a term that the contractor should warn the architect of defects in his design.

In the *University Court of Glasgow* case it was also alleged that the contractor was under a tortious duty to warn of design defects. However, in view of the developments in the law of tort since the *EDAC* decision, which excluded the possibility of recovering economic loss for negligence except in very special circumstances, Judge Bowsher held that any duty on the contractor was confined to an obligation to avoid acts or omissions which resulted in personal injury or physical damage to property other than the building itself. Thus, in most cases, and in the *University of Glasgow* case itself, a duty to warn in tort would have no application.

Although the trend of the more recent cases has been away from imposing a duty to warn, sub-contractors should nevertheless be aware that the law on the issue is not fully settled, and so they may become caught up in cases about defective design where it is clear that the main responsibility lies with another party. This can involve them in enormous risk because of the legal rules on joint and several liability, which provide that where two or more breaches of duty (either breaches of contract or negligence) cause someone to suffer a single injury (for example a defective building) the injured party may sue all or any of those who are in breach of duty for the full amount of his loss.

Although anyone who is sued has a statutory right under the Civil Liability (Contribution) Act 1978 to recover a contribution to the damages he has had to pay from the other responsible parties, this does not affect the position of the injured party, who may recover his loss in full from whoever he chooses. The significance of the duty to warn in this context is that it may involve the sub-contractor in liability where he is perhaps only 10 per cent to blame for a design problem, as the lion's share of the responsibility lies with the person who supplied the defective design. However, if that person is insolvent or uninsured, the injured party may be able to recover the whole of his loss from the sub-contractor, whose right of contribution against the real designer would be worthless.

Non-contractual liability for design

Quite apart from any liability which may be created by a contract, a sub-contractor may also be legally liable for design due to the operation of the law of tort, or because a duty is imposed upon him by statute.

Where the injured party has a range of remedies, suing in contract is likely to be the favoured option. There are two reasons for this: firstly it is generally easier to prove breach of contract than it is to establish either negligence or a breach of statutory duty; secondly the amount of damages which can be recovered for breach of contract is often greater.

However, sometimes an action in contract will not be available to the person who has suffered from the defective design, generally because he does not have a contract with the person responsible for it. In such circumstances, he will be forced to consider his other remedies, which are as follows.

Liability in negligence

In principle, a sub-contractor carrying out design can be sued by any person to whom he owes a duty of care, provided that person has suffered a recoverable type of loss. But decisions of the courts in cases such as *Murphy v. Brentwood* [1990] have made it difficult to obtain a remedy in negligence by limiting the types of loss which a plaintiff can recover.

As the law currently stands, in order to recover damages for negligence, the plaintiff must have suffered:

- death or personal injury; *and/or*
- physical damage to other property (ie property other than the product of the negligence).

What is often described as 'pure' economic loss will not be recoverable. In the context of building contracts, this excludes the majority of the losses which an injured party will be seeking to claim, namely:

- the costs of remedying the negligently executed work or design; *and*
- consequential costs such as loss of rent or loss of profit while the building cannot be used, either because the defects are so serious it is uninhabitable or because remedial work is being carried out.

In some circumstances, economic losses such as loss of profit may be recoverable in negligence if they are directly consequent on personal injury or damage to other property. However, none of the cases where damages have been awarded for this type of loss relate to building contracts, and therefore it would be unwise to assume that this principle has a wide application.

It is because of these restrictions on the type of loss which can be recovered in negligence that establishing a contractual link with the designers, main contractors and sub-contractors by means of a collateral warranty will be so important to any purchasers or tenant of the building.

Negligent misstatement

There is one very important exception to the rule that economic loss is irrecoverable, which relates directly to liability for design. This is where there has been a negligent misstatement.

If someone negligently gives bad advice or incorrect information, the recipient may have a remedy for negligent misstatement, provided he can establish that he had a special relationship with the person who advised him.

This rule was laid down in the case of *Hedley Byrne* v. *Heller & Partners* [1964] AC 464, in which the House of Lords held that if one party seeks information from another in circumstances where a 'special relationship' exists, a duty to exercise reasonable skill and care will be imposed upon the party giving the information. If he fails to do so, he will be liable to the other party in damages, which will include damages for purely financial losses.

A 'special relationship' will exist where, in a particular sphere of activity, a person is so placed that others can reasonably rely on his skill and judgment, or on his ability to make careful enquiries, and that person gives information or advice to someone he knows will place reliance upon it.

It is easy to see how such a relationship could arise between a sub-contractor and the employer. Many sub-contractors are uniquely placed in terms of their knowledge of the product which they supply, and therefore if they advise the employer (or indeed his professional team) on the selection of, for example, a particular fire alarm or cladding system, or they participate in design by suggesting how their product or system could be integrated into the building, they could be liable for negligent misstatement if they get it wrong.

Liability for negligent misstatement is not dependant on payment for the advice or information, so even the information so willingly given to a potential client who relies upon it but does not employ that sub-contractor to actually do the work can create legal liability.

Statutory liability – Defective Premises Act 1972

A sub-contractor may be under a statutory duty under the Defective Premises Act 1972 if he participates in the design of any building which could be described as a dwelling – for example a private house or a block of flats.

Section 1(1) of the 1972 Act provides that

> 'a person taking on work [**which includes the provision of design**] for or in connection with the provision of a dwelling . . . owes a duty to every person who acquires an interest in the dwelling . . . to see that the work which he takes on is done in a workmanlike or, as the case may be, professional manner, with proper materials and so that as regards that work the dwelling will be fit for habitation when completed'.

As is evident from Section 1(1), this Act opens up the sub-contractor to claims from a wide range of people – in particular future purchasers and tenants of the property. The duty he owes to them is a strict one in the sense that he must 'see that the work which he has taken on is done properly', so it will be no defence for him to show that he has exercised reasonable skill and care.

There is, however, an important limiting factor in that, in order to establish liability under the Act, a plaintiff must show that the dwelling is unfit for habitation. This will exclude many types of defect from its scope. A heating system which does not function perfectly, for example, or windows which let in draughts are undoubtedly defects, but are unlikely to render a dwelling unfit for habitation. Problems with the foundations, or perhaps the roof, are much more likely to fall into that category.

There is often confusion as to whether the Defective Premises Act applies to a house or flat which is covered by the National Housebuilding Council's (NHBC) guarantee. This is because Section 2 of the Act prevents an action being brought under the Act where the dwelling is covered by 'an approved scheme'. Until 1979, the NHBC schemes were approved schemes and therefore a significant proportion of dwellings built up to 1979 were not covered by the Act. Since 1979, there have been no 'approved schemes', and so the Act applies even where the dwelling is covered by an NHBC guarantee or similar scheme.

The limitation period in respect of actions under this Act starts to run when the dwelling is completed (**not** when the sub-contractor has completed his work, which may be considerably earlier) and runs for six years. If rectification work is done after the dwelling is complete, the six year limitation period starts to run again in respect of that section of the work which has been rectified.

Copyright in drawings, plans and buildings

English law automatically vests copyright in the author of any 'original artistic works' which include drawings and plans and indeed the building itself. The requirement for originality is not applied stringently, but does mean that a pure copy of someone else's design will not attract copyright protection. However, even if a design is simple and incorporates standard modules or basic features it may still be an original artistic work.

If an artistic work is created by an employee in the course of his employment, ownership of the copyright vests in his employer unless they have expressly agreed otherwise.

These two rules mean that any sub-contractor who designs a substantial part of his works is therefore likely to hold the copyright in that design.

Copyright holders are protected by the Copyright, Designs and Patents Act 1988, and can claim an injunction or damages for any breach of copyright. This does not mean that the sub-contractor's design cannot be used, only that any user must have a licence to do so.

The law will generally imply a licence in favour of any person who has commissioned and paid for the design, and therefore if the sub-contractor has been employed to design as well as instal the sub-contract works, the employer, or possibly the main contractor, will acquire a non-exclusive licence to the design. However, where design is 'hidden' in the sense that it is not obvious that the sub-contractor is being employed to design although that is what he is actually doing, and there is no evidence that he is receiving payment for it, an express license may be necessary if copyright problems are to be avoided.

The subject is often dealt with expressly in non-standard employer sub-contractor warranties, where the sub-contractor is required to grant a royalty free non-exclusive licence to the beneficiary of the warranty, usually a purchaser or tenant. However, the use of the design is usually restricted to the maintenance, repair and possibly extension of that building and, where the sub-contractor has been sharp in the warranty negotiations, will be conditional upon full payment for the design.

Copyright protection under the Act lasts for fifty years from the death of the originator.

Professional indemnity insurance

None of the JCT standard forms or their related sub-contracts or employer/sub-contractor warranties discussed above require the designing contractor or sub-contractor to carry professional indemnity insurance. However, JCT Practice Note CD/1A, which relates to the Design and Build Form 1981, acknowledges that such insurance may be required by the employer, and in practice this is almost invariably the case. This can create real problems for the sub-contractor, who may find that he cannot obtain professional indemnity insurance because none of his designers have any formal qualifications or, if he can buy it, that it is prohibitively expensive.

For a full discussion on professional indemnity insurance, see Chapter 10.

5 Workmanship and Materials; Defects Liability

Sub-contractors must supply goods, materials and workmanship which are in accordance with the express and implied terms of their sub-contracts. If they fail to do so, they will be in breach of contract and the main contractor will be entitled to recover damages for any loss he suffers as a result. That basic principle sounds (and is) perfectly straightforward, but it has given rise to some complex contractual rules regarding liability for defects in the sub-contract works.

This chapter summarises the express terms relating to quality of work and materials in the JCT related sub-contracts, and the additional terms which will be implied into those sub-contracts by the general law. It also describes the powers of the main contractor and architect when defects arise and considers the effect of the final certificate on the sub-contractor's liability.

There is a common misconception that liability for defects in building work comes to an end on the expiry of the defects liability period specified in the relevant contract. This is not the case: the sub-contractor will remain legally responsible for defects in his work, provided they were due to his negligence or breach of contract, until the expiry of the relevant limitation period. The rules on limitation are therefore set out at the end of this chapter.

Express terms as to quality

The sub-contracts for use with the different JCT standard forms lay down similar standards as to quality of goods, materials and workmanship.

The basic requirement which they all contain is that both the completed sub-contract works and the goods, materials and workmanship which they comprise must comply with the sub-contract documents. The sub-contract documents are defined as the sub-contract itself (that is the articles of agreement, conditions and appendix) and the numbered documents, which are the documents describing the sub-contract work such as the drawings, specification and bill of quantities. They are the documents on which the sub-contractor will have based his tender.

If the numbered documents include a specification, that will usually be where most of the requirements as to quality will be set out, although there may also be relevant information on drawings and in any bill of quantities. This means that there is a potential for conflicts to arise − for example, the specification may refer to a particular type of guttering, but a note on the drawings specifies something entirely different. When the sub-contractor finds such a discrepancy (he is not obliged to look for them in advance) he is under a contractual obligation to notify the contractor and to ask for directions. Compliance with those directions will be treated as a variation.

In addition the sub-contractor's obligation to comply with the sub-contract documents, he may also have to satisfy a contractual requirement that materials, goods and workmanship are to the architect's reasonable satisfaction.

For example, clause 4.1.2 of DOM/1 states:

'All materials and goods shall, so far as procurable, be of the kinds and standards described in the Sub-Contract Documents provided that where and to the extent that approval of the quality and standards of materials and goods is a matter for the opinion of the Architect such quality and standards shall be to the reasonable satisfaction of the Architect.'

Although clause 4.1.2 makes it fairly clear that not all materials and goods are 'a matter for the opinion of the Architect', its wording gives the parties little assistance in determining exactly what will fall into that category.

The lawyer's answer to such a question would be that it will be a question of construction of the sub-contract documents in every case. That is not much help to anyone who is confronted with a practical problem, but fortunately it is possible to illustrate how the clause will operate by some examples.

If the sub-contract documents clearly reserve a matter to the opinion of the architect, for example 'plastering to be finished to the architect's approval', there is no doubt that the standard of workmanship must be to the architect's reasonable satisfaction. Conversely, if the specification only lays down an objective standard with which an item of equipment must comply, for example it refers to the relevant British Standard, it is also reasonably clear that there will be no additional requirement for that equipment to be to the architect's reasonable satisfaction.

Between those two extremes, the matter will be open to argument. Until recently, it was thought that a clear intention in the numbered documents would be required in order to bring any particular thing within the scope of the architect's approval. However, when the issue was considered by the Courts in the recent case of *Colbart Ltd* v. *H. Kumar* (1992) 59 BLR 89, they reached a different conclusion.

In that case, the court was required to decide whether the issue of a final certificate under the Intermediate Form 1984 prevented the employer from refusing to pay certified sums to the contractor on the grounds that certain work was defective. Clause 4.7 of IFC 84, in common with most JCT standard forms, provides that the final certificate is conclusive proof that, where the quality of materials or the standard of workmanship is a matter for the opinion of the architect, they are to his reasonable satisfaction. It will not, however, have that effect in respect of any matter which is the subject of legal proceedings commenced within 28 days of the date the final certificate was issued.

The judge in *Colbart* v. *Kumar* therefore had to decide precisely what was a matter for the opinion of the architect. He held that it was not restricted to materials and workmanship which were **expressly** reserved by the contract (for example in the specification) as a matter for the opinion of the architect, but included all materials and workmanship the approval of which was **inherently** a matter for the opinion of the architect. Unfortunately, he gave little guidance as to what those matters might be beyond stating that it would be a question of fact and degree in each case.

Although the defects on which the employer was attempting to rely in *Colbart* v. *Kumar* are not listed in the law report, it is understood that they were of a relatively minor nature, and were not limited to aesthetic matters, such as the quality of finishes, which might normally be considered to be the only matters which are inherently a matter for the architect's opinion. Nevertheless, the judge found that the defects in that case were inherently a matter for the opinion of the architect, and therefore the employer was bound to pay the sums certified.

That decision has caused some consternation among architects, who are now less willing to issue final certificates for fear that it will deprive their clients of a remedy in the event of the discovery of defective work. It is understood that the JCT was discussing in 1994 whether to amend its standard forms to provide that only those matters expressly stated in the contract documents to be a matter for the opinion of the architect are required to be to his reasonable satisfaction, and similarly to restrict the conclusive effect of the final certificate.

The decision in *Colbart* v. *Kumar* was distinguished in a subsequent case in the Official Referees' Court, *Darlington Borough Council* v. *Wiltshier Northern Ltd* (1993) unreported. The *Darlington* case concerned similar provisions in the 1963 Edition of the JCT standard form, and on the grounds of the slightly different structure of the relevant clauses in JCT 63, it was held that the final certificate was final and conclusive only in relation to those matters which are expressly stated in the contract documents to be a matter for the architect's opinion. As the relevant clauses in JCT 80 have more in common with those in JCT 63 than those in IFC 84, it could be argued that the decision in *Colbart* v. *Kumar* is of limited application.

It is a widely held belief in the construction industry that the issue of the final certificate relieves the contractor of liability for all defects in his work, particularly any defects which were patent (i.e. obvious) at the date the final certificate was issued. As *Colbart* v. *Kumar* and the *Darlington* case illustrate, this is totally incorrect. The misconception probably stems from pre-1976 versions of the JCT forms, which did provide for the final certificate to be conclusive proof that all work and materials were in accordance with the contract.

Implied terms as to quality

An implied term is one which the general law will treat as being incorporated into the contract, unless the parties have expressly agreed to exclude it. In contracts as detailed as those published by the JCT, there is relatively little scope for implied terms because there are express terms dealing with most matters. However, there are certain terms which will usually be implied either by statute or by the common law into the sub-contracts considered in this book.

A building contract is a contract for the supply of services, and therefore falls within the scope of the Supply of Goods and Services Act 1982. Section 13 of that Act implies a term that the supplier of a service will carry it out with reasonable care and skill.

In addition to this statutory implied term, the courts will imply certain terms into a building contract unless the parties have expressed an intention to exclude or modify them. In the event of a dispute as to whether one of these usual terms is to be implied, the burden of proof will fall upon the person alleging that it is not incorporated into the contract.

The terms to be implied into building contracts in relation to quality of materials were considered by the House of Lords in two cases decided in 1969: *Young & Marten* v. *McManus Childs* [1969] 1 AC 454 and *Gloucestershire County Council* v. *Richardson* [1968] 2 ALL ER 1181. The basic rule is that there are implied terms that materials supplied by the contractor or sub-contractor will be:

(i)   of good or 'merchantable' quality; *and*
(ii)   reasonably fit for the purpose for which they will be used.

unless the express terms of the contract or the surrounding circumstances show that the parties intended to exclude either or both such terms.

Proving that the parties intended to exclude a term that the materials supplied would be of merchantable quality will not be an easy task for the sub-contractor. For example, the fact that defects may be latent (i.e. hidden), and that no amount of investigation at the time of supply would have revealed the defect, will not assist him. He may, however, be more successful if he can show that all of the details of the contract of supply, such as price, design and quality, were fixed by the employer (or his agent) or the main contractor, and that he had no right to object to the nomination of the supplier.

This was the position in *Gloucestershire County Council* v. *Richardson*, where a main contractor was instructed to accept a quotation for the supply of certain concrete columns. The quotation contained a term that the supplier's liability for defective goods was limited to free replacement, and under the terms of the main contract, the contractor had no right to object to contracting with that supplier. The House of Lords held that, in those circumstances, the contractor was not liable to the employer for latent defects due to bad manufacture.

It is a much simpler proposition for a sub-contractor to prove that there should be no implied term that materials must be fit for their purpose, provided they have been selected by someone other than the sub-contractor. Generally, a term as to the suitability of goods and materials for any particular application will be implied only where the following criteria have been fulfilled:

(i)   the employer or main contractor has made known to the sub-contractor the particular purpose for which the goods or materials are required; *and*

(ii) the employer or main contractor relied on the sub-contractor's skill and judgment in selecting the goods or materials.

Whether the main contractor or employer relied on the sub-contractor's skill and judgment will be a question of fact in each case. If an item is clearly specified in the numbered documents it will be virtually impossible to prove reliance, and the sub-contractor's only obligation will be to supply a good example of the specified item. Equally, if the contractor has an unfettered right to select goods or materials, there will be a strong inference that he warrants they will be fit for their purpose. Often, the situation will not be so clearcut, as a range of goods may be specified from which the sub-contractor may choose. In those circumstances, it will be necessary to examine the precise circumstances to ascertain the extent of any reliance on the sub-contractor's skill and judgment.

A term that materials must be of good or merchantable quality will be implied notwithstanding the fact that the employer or main contractor may have chosen the materials or nominated who should supply them. For example, in *Young & Marten* v. *McManus Childs*, the employer's agent relied on his own skill and judgment in choosing 'Somerset 13' tiles to be fixed by a sub-contractor. The tiles appeared to be sound, but within twelve months of being fixed some of them began to disintegrate. The House of Lords held that although the warranty as to fitness for purpose was excluded because the employer's agent had selected the type of tile to be used, the warranty as to quality was not, and therefore the contractor was liable.

It is therefore no defence for a sub-contractor who has supplied defective materials to claim that he is not liable because they were selected by someone else; he remains bound to supply a good example of the thing specified.

In certain circumstances, a term may be implied into the sub-contract that the whole of the sub-contract works, rather than simply particular materials and goods, are to be fit for a particular purpose. As this generally arises only in the context of contractor designed works, it is discussed in Chapter 4.

It is also likely that a term will be implied into all building contracts that the contractor (or sub-contractor), must carry out his work with all proper skill and care, although it is an open question whether this means that defects occurring before practical completion which are remedied by the sub-contractor of his own volition constitute an actionable breach of contract. This is discussed further under the next sub-heading.

Defects discovered before practical completion

There is some debate as to whether a defect arising during the progress of the sub-contract works constitutes a breach of contract which, even if satisfactorily remedied, entitles the main contractor to recover damages at common law. The point has been discussed in a number of cases, notably in *Hosier & Dickinson v. Kaye* [1970] 1 WLR 1611 where it was suggested that such a defect was merely a 'temporary disconformity' and therefore did not constitute a breach entitling the other party to recover damages.

This view is not shared by many leading commentators, and some doubt has been cast upon it in subsequent cases, e.g. *Lintest Builders Ltd v. Roberts* (1978) 10 BLR 120. One reason for doubting the 'temporary disconformity' approach is that once goods and materials are attached to land they become the property of the owner of that land (usually the employer). Further, the JCT forms make it clear that the express powers given to the architect or main contractor on the discovery of a defect are without prejudice to any other rights and remedies the other party to the contract may possess, so any rights which do exist at common law will be preserved.

Bearing in mind the disruption which can be caused by remedying a defect at a late stage in the works, it is suggested that it would be prudent for any contractor or sub-contractor who has executed defective work to assume that damages for breach may well be payable and therefore to ensure any costs consequent on the defect are kept to a minimum.

In addition to any claim for damages at common law, the JCT standard forms of main contract and their related sub-contracts give the architect (or the main contractor under the domestic sub-contracts) a wide range of powers which may be exercised when defective work, materials or goods are discovered before practical completion. Many of these powers are relatively new and untried, as they were not introduced into JCT 80 until 1988. They are all capable of involving the contractor or sub-contractor responsible for the defect in considerable expense, particularly those relating to opening up and testing of work after the discovery of an initial defect, and therefore repay careful study.

Each of the JCT standard forms contains different provisions on this subject, so they have been dealt with separately in the next section. In addition, because it is necessary to appreciate the powers of the architect under the main contract in order to understand the position at sub-contract level, the main form rules have also been summarised in each case.

JCT 80

The architect's powers on the discovery of any work, materials or goods which are not in accordance with the contract are set out in clause 8.4.

Firstly, he is entitled to require the contractor to remove any defective work, materials or goods, whereupon the contractor will be bound to re-execute the work or re-supply goods and materials which are in accordance with the contract. Prior to Amendment 5 to JCT 80, which was issued in 1988, this was the only kind of instruction relating to defective work which the architect was expressly empowered to issue.

Now, however, the architect also has an express power to allow defective work, materials or goods to remain, in which case an appropriate deduction can be made from the contract sum. Before deciding to exercise this power, the architect must consult with the contractor, and must also obtain the agreement of the employer. The architect may wish to allow defective work to remain if the defects are relatively minor, such as scratches on paintwork or defective bulbs on a control panel. Alternatively, he may do so because the employer urgently requires the building on a specific date, and liquidated damages will not fully compensate him for any delay. This may occur where, for example, a local authority requires a school to be open at the beginning of a new academic year.

The precise wording of clause 8.4.2 under which the architect acquires this power is important: he does not actually issue an instruction that the work should be retained, he merely 'allows' it to remain. If the contractor were to object to the retaining of such work, perhaps on the grounds of aesthetics, or even more importantly, safety, it is debatable whether the architect could insist on the work remaining without issuing an instruction to that effect, which could then open the door to a claim by the contractor that the instruction amounted to a variation, and might also exonerate him from any consequences of the defect remaining unrectified.

Under clause 8.4.3, the architect may instruct such variations as are reasonably necessary as a consequence of either requiring the removal of defective work, materials and goods, or allowing them to remain. Such variations will be, in effect, free to the employer because the contractor will not be entitled to the cost of complying with such instructions, nor will he receive an extension of time or any loss and expense for any delay it might cause. Again, the architect must consult with the contractor before exercising this power, and although this does not bind him to take account of any comments the contractor makes, it will allow the contractor to suggest cheaper or more efficient ways of overcoming the problem.

Clause 8.5 gives the architect a similar power to issue 'free' variation instructions where the contractor has failed to carry out work in a proper and workmanlike manner.

The architect's powers to instruct opening up or testing once an initial defect has been discovered were significantly increased by Amendment 5. The architect has always had a general power under JCT 80 to instruct the contractor to open up for inspection any work covered up, or to arrange for or carry out any test. Unless that testing was required by the contract documents or it revealed a defect in the work tested, the contractor would be entitled to recover the costs of carrying out the opening up or testing from the employer, together with an extension of time and loss and expense for any delay which it caused. If defects were discovered, the opening up or testing was entirely at the contractor's cost, and he was not entitled to an extension of time or to any loss and expense. (This basic power to instruct opening up or testing has been retained in clause 8.3 of JCT 80.)

Although there was an element of rough justice in this approach, it seemed to be satisfactory until the problem of defects in work which was repeated throughout the building was raised in the JCT. An illustration of that problem would be a contract for the construction of a block of a hundred flats, each with a balcony. Suppose a test is carried out on one of the balconies, and it proves to be unsafe. That test would be paid for by the contractor. But, it was argued, a prudent architect would not stop there, but would quite properly require tests to be carried out on at least some of the other balconies. If those subsequent tests revealed no further defects, the contractor would be entitled to payment in full for those tests, and to an extension of time and loss and expense. This would seem unfair to the employer, as the only reason the tests were ordered was because of the contractor's defective work on the first balcony tested.

To accommodate this eventuality, special rules on opening up or testing after a defect has been discovered have been included in clause 8.4.4. They give the architect the right to instruct such further opening up and testing as is reasonable in all the circumstances to establish to his reasonable satisfaction the likelihood or extent of further, similar defects. Providing the instruction is reasonable, that testing will be at no cost to the employer, even if it demonstrates that all of the work, materials and goods are in accordance with the contract. If no further defects are revealed, he will get an extension of time, but no loss and expense.

These arrangements may tempt architects to issue sweeping instructions for further testing when defects are discovered, on the basis that they will not be an additional cost to their client, but will reassure him that there are no similar defects and also ensure that the architect himself cannot be accused of negligence by failing to instruct sufficient testing on the discovery of a defect. This is a temptation which should be resisted, as the architect's powers in these circumstances are actually quite restricted. In particular, he should take into account the following matters:

- when issuing instructions for further opening up or testing, he must have due regard to a code of practice in JCT 80 which was drafted to ensure the fair and reasonable operation of clause 8.4.4;
- the testing instructed should be only that which is necessary to establish the likelihood or extent of further, similar defects, so he may not use the discovery of a minor defect as an excuse to order testing which is 'free' under clause 8.4.4 of unrelated work, materials or goods;
- the contractor will obtain an extension of time in the event that the further tests do not reveal any more defects, and therefore substantial further opening up and testing could delay the completion of the works in circumstances where the employer would not be entitled to recover liquidated damages;
- if the architect's instruction is unreasonable, it will be treated as having been issued under clause 8.3, which is the ordinary clause on opening up and testing, and therefore if no defects are revealed, the contractor will be entitled to the costs of the testing and to loss and expense, as well as to an extension of time.

Any dispute or difference on an architect's instruction under any part of clause 8.4 is subject to immediate arbitration, so a contractor wishing to challenge an architect's decision on the amount of opening up or testing required after an initial defect had been discovered would not have to wait until practical completion to do so.

The code of practice which was mentioned earlier is of vital importance to the proper operation of clause 8.4.4, and would almost certainly be taken into account by any arbitrator who was called upon to decide whether an architect's instruction in this regard was reasonable. It states that the architect and contractor should try to agree the amount and method of opening up or testing, but if that is not possible, the architect should take a number of matters points into account when issuing his instruction. These include:

- the need to discover whether the defect is such that rigorous testing of similar elements must take place or whether it may indicate an inherent weakness which requires selective testing;
- the significance of the defect;
- the consequence of any further similar defect on the safety of the building;
- the standard of the contractor's supervision of the works;
- the reason for the defect where it has been established;
- current recognised testing procedures;
- the practicability of progressive testing to establish whether any further similar defects are likely;

- if alternative testing methods are available, the time and cost consequences of each of them;
- any proposals of the contractor.

The code therefore makes it clear that the architect must not act capriciously
and order wholesale opening up and testing because of the discovery of one
minor defect. It also encourages the contractor to be proactive, by providing
information as to the reason for the defect and making his own proposals as
to how it can be demonstrated that there are no further similar defects in the
works.

If operated sensibly, these provisions should not create a major problem for
either architects or contractors. A reasonable architect would, as a matter of
course, discuss the amount of further testing required with the contractor,
who should be equally anxious to demonstrate that the failure of his work was
a 'one-off', and does not demonstrate any inherent weakness. An unreasonable architect could, however, make life very difficult for the contractor,
particularly in relation to work where a small number of defects are to be
statistically expected, such as welding. In these circumstances the code of
practice will be very important to the contractor in helping him to demonstrate that any further testing should be on a progressive or selective basis.

Nominated sub-contract NSC/C

Clause 3.4 of NSC/C provides that any instruction of the architect under
clause 8.4 or 8.5 affecting the nominated sub-contract works must be passed
on to the nominated sub-contractor, whose rights and obligations will be
exactly the same as those of the main contractor under JCT 80.

The main form provisions require that, wherever the architect is required to
consult with the contractor, for example before deciding to allow defective
work to remain, the main contractor must immediately consult with any
relevant nominated sub-contractor. NSC/C also binds the contractor to consult with the nominated sub-contractor where appropriate, and requires him
to report the outcome of that consultation in writing to the architect, with a
copy to the nominated sub-contractor. The sub-contractor is therefore able to
enforce his right to be consulted if necessary, and can also ensure that his
views have been accurately reported to the architect.

The main contractor is given no express powers in NSC/C to issue his own
directions to the nominated sub-contractor on the matters covered by
clauses 8.4 and 8.5 of the main contract; he can only pass on instructions
from the architect. It is likely, however, that he has an implied right to issue
directions requiring the removal and re-execution of defective work, because
the cases on the main forms indicate that such a power would exist in those
contracts even if there were no express provisions to that effect.

In addition to the stepping down of the main contract clauses dealing with defects discovered prior to practical completion, NSC/C also contains some quite complex arrangements which deal with the effect on the nominated sub-contract work of instructions relating to defective work executed by others. This might occur where, for example, further testing is required of services contained in a ceiling void which require the ceiling contractor to take down and subsequently replace part of his work.

Clause 3.8 of NSC/C provides that where any instruction issued under clause 8.4 or 8.5 necessarily results in work properly executed by the nominated sub-contractor being taken down or re-executed, or in goods and materials being re-fixed or re-supplied, the main contractor may issue directions to the sub-contractor requiring him to carry out that taking down, etc. provided the sub-contract works have not yet reached practical completion.

The nominated sub-contractor will be entitled to be paid for the work he carries out pursuant to that direction, but it will not be part of the sums certified as due to him in architect's certificates. Such work will be treated as a domestic matter between the main contractor and his sub-contractors, and therefore the amount to be paid will simply be calculated on the basis of what NSC/C describes as a 'fair valuation', which will presumably be carried out by the main contractor. Payment for the work should be made within 14 days of the end of the month during which the relevant work was carried out. The sub-contractor will also be entitled to an extension of time and to loss and expense for any delay caused by complying with such directions.

If the taking down, etc. work is required after the nominated sub-contract works have reached practical completion, the main contractor has no power to force the nominated sub-contractor to carry it out. There is nothing to prevent him reaching an agreement with the nominated sub-contractor for him to do so, but if he cannot, or if he does not wish to, the architect is bound to nominate a new sub-contractor to carry out the taking down and re-fixing, etc.

NSC/C also contains an indemnity in clause 3.9 from the sub-contractor to the main contractor against any costs the main contractor may incur due to the operation of the clauses described above in other nominated sub-contracts due to defective work carried out by that nominated sub-contractor.

Domestic sub-contract DOM/1

The clauses in the standard form of domestic sub-contract, DOM/1, are different from those in NSC/C in one vital respect: they give the main contractor his own powers to issue instructions relating to defects arising before practical completion; he is not restricted to passing on architect's instructions.

The range of powers available to the main contractor is set out in clause 4.3 of DOM/1.

One of the actions he can take is to direct the removal or rectification of any defective work, which is described in the sub-contract as 'non-complying work'. If he is issuing the direction on his own initiative, rather than simply passing on an architect's instruction issued under the main contract, he must have regard to a code of practice appended to the DOM/1 conditions (designated Code of Practice A to distinguish it from the code of practice relating to opening up and testing).

The purpose of the code is to assist him in deciding whether to direct removal or rectification of the non-complying work. The matters he is required to take into account include the relative cost of removal and rectification, the amount of time each of them would take and the significance of the defect. The code also emphasises the desirability of the main contractor and domestic sub-contractor reaching agreement on the way in which the non-complying work should be handled, and it will be in the sub-contractor's own interests to make sensible proposals in this regard.

The remainder of the main contractor's powers in relation to defective work discovered before practical completion are the same as those of the architect under clauses 8.4 and 8.5 of JCT 80, subject to two important restrictions.

Firstly, the contractor may not direct that non-complying work should be allowed to remain, or to vary the works free of charge as a consequence of such allowance: these matters were obviously felt to be so important that they should be reserved to the architect. Secondly, where materials, goods or workmanship are required to be to the architect's reasonable satisfaction, the main contractor has no power to order their rectification or removal or further opening up or testing on his own initiative: he may only pass on instructions from the architect.

Domestic sub-contractors who are concerned at the prospect of the main contractor having a free hand to require further opening up or testing of the other areas of their work should take comfort from the two additional matters in Code of Practice B to which the main contractor must have regard before issuing his directions. They are:

- the nature and extent of any opening up or testing instructed by the architect under the main contract which affects the domestic sub-contract works; *and*
- any decision of the architect under the main form not to instruct opening up or testing which affects the domestic sub-contract works.

This will make it relatively difficult for the main contractor to issue a 'reasonable' direction as required by DOM/1 which requires testing additional to that required by the architect under the main contract.

Provisions similar to those in NSC/C regarding taking down and refixing work to allow others to test their defective work are also included in DOM/1.

Intermediate Form 1984

Compared to his powers under JCT 80, the architect's right to issue instructions under IFC 84 on the discovery of defective work before practical completion are somewhat limited.

He is entitled to issue instructions requiring the removal from site of any defective work, materials or goods under clause 3.14. He may also issue any instructions which are necessary as a result of work not being carried out in a proper and workmanlike manner, with which the contractor must comply at no cost to the employer. However, he is not entitled to instruct 'free' variations in any other circumstances, and he has no right to allow defective work to remain.

In addition to the usual power to order any opening up or testing not required by the contract documents on the basis that if there are defects the contractor pays, and if there are none the employer pays, the architect has special powers in this regard if defective work has already been discovered, although the IFC 84 provisions are different, and perhaps easier to operate, than those in JCT 80.

Under clause 3.3.1 of IFC 84, as soon as a defect is discovered, the contractor must make proposals in writing to the architect as to the action he proposes to take at no cost to the employer to establish that there are no further similar defects. If the contractor fails to provide that statement within seven days of the defect being discovered, or if the architect is not satisfied with what the main contractor is suggesting, he may instruct such further opening up or testing as he considers necessary to establish there are no further similar failures. He may take this action immediately if he is unable to wait for the contractor's proposals because of safety considerations or statutory obligations.

The contractor must comply with any such directions, and at his own cost, regardless of the outcome of the tests, although he will be entitled to an extension of time if all of the work tested is in accordance with the contract.

The contractor may object to the amount of testing instructed by the architect under clause 3.3.1, provided he does so in writing within ten days of receiving the instruction. However, this does not affect his obligation to comply with it, and if the architect refuses to withdraw or modify the instruction, the contractor's only remedy will be to challenge the instruction as unreasonable in immediate arbitration.

As there is no code of practice in IFC 84 to which the architect must have regard before issuing instructions on opening up and testing, contractors may feel it is more difficult to mount such a challenge of an instruction under IFC 84 than under JCT 80, although similar matters to those listed in the code are likely to be taken into account by any arbitrator appointed under IFC 84 called upon to determine the reasonableness or otherwise of such an instruction.

Named sub-contract NAM/SC

The main form provisions on these issues are stepped down into NAM/SC by clause 5. Unlike NSC/C, this sub-contract gives the main contractor his own, unfettered power to require the removal of defective work and to order opening up or testing after the discovery of an initial defect.

The time periods at sub-contract level are, of course, slightly shorter: the sub-contractor's proposals must be submitted to the main contractor within five days of the discovery of the defect, and he has only seven days to object to any direction requiring further opening up or testing.

The provisions which have been included in NSC/C and DOM/1 regarding the need to take down and re-fix work because of defects in the work of other trades have not been included in NAM/SC, but it is suggested that their omission will not cause any great difficulty, as in practice any direction from the main contractor requiring such work to be carried out will amount to a variation under NAM/SC, and the sub-contractor will therefore be entitled to recover the full cost of complying with it, as well as an extension of time and loss and expense. Those costs will be recoverable from the sub-contractor who executed the defective work under the standard indemnity in clause 6 of NAM/SC against losses due to the sub-contractor's act, omission or default.

The contractor may also issue such directions as are reasonably necessary as a consequence of the sub-contractor failing to carry out the sub-contract works in a good and workmanlike manner, and the sub-contractor must comply with those instructions at his own expense.

Domestic sub-contract IN/SC

The contractor's powers under the standard form of domestic sub-contract for use with IFC 84 are exactly the same as those under the named sub-contract NAM/SC.

JCT with Contractor's Design Form 1981

The employer has broadly the same powers under JCT 81 as the architect has under JCT 80 in regard to issuing instructions following the discovery of defective work and materials before practical completion, subject to the same protections for the contractor.

The only material difference is that the employer has no express power to allow defective work to remain, as it was felt that this would be inappropriate in a contract where the employer may have no professional advice and the contractor was, in part at least, responsible for the design of his works.

Domestic sub-contract DOM/2

The position of the parties under DOM/2 is exactly the same as their position under DOM/1, save that as there is no power in the main contract to allow defective work to remain, there is no such power in the sub-contract.

JCT Management Contract 1987

Clause 3.10 entitles the architect to instruct the management contractor to secure the carrying out by works contractors of opening up or testing, which will be at no cost to the employer if it reveals defects in the work tested.

Provisions similar to those included in JCT 80 by Amendment 5 have not, at the time of writing in February 1994, been included in either the Management Contract or the Works Contract, but it is understood that the JCT intends to publish an amendment including such provisions during 1994.

Defects appearing after practical completion

All of the JCT standard forms and their related sub-contracts provide for a defects liability period, during which the contractor or sub-contractor is obliged to rectify any defects which appear and which are notified to him by the architect (or, in the case of the sub-contracts, by the main contractor). This arrangement has advantages for both sides: it reassures the employer that any problems appearing during the early life of the building will be resolved, and it allows the contractor or sub-contractor to remedy defects himself, which will usually be cheaper than reimbursing the employer the costs of someone else doing so. In addition, the contracting side will have every incentive to respond to notifications of defective work during this period, as the employer will still be holding half of the retention fund.

All of the JCT forms contain similar provisions relating to rectifying defects during the defects liability period, and they are therefore commented upon collectively.

JCT main contracts

The defects liability period will commence on the date of practical completion of the works, and will run for the length of time specified in the appendix, which will usually be six months.

During that time, the architect (or the employer under JCT 81) may issue instructions requiring the contractor to make good defects appearing within the defects liability period at his own cost, providing the defect is due to materials or workmanship not being in accordance with the contract, or to frost occurring before practical completion. Defects due to other causes, such as misuse by the occupier of the building or the effect of weather on the building after practical completion, do not entitle the architect to insist on rectification by the contractor.

During the defects liability period, the architect has the option of instructing that certain defects should not be made good, but to make an appropriate deduction from the contract sum in respect of those defects. This power, unlike the similar right to allow defective work to remain before practical completion, is not restricted by any requirement to consult the contractor before it is exercised.

It is important to appreciate that, strictly speaking, the architect is entitled to require the contractor to rectify only those defects **appearing** after practical completion – any defects which were patent (i.e. obvious) at the date of practical completion are not covered by the express terms of the contract and should not be included in any list of defects issued by the architect under the clauses relating to defects liability.

This slightly pedantic distinction is generally ignored by architects, who habitually send a list of defects to the contractor at the same time as they certify practical completion, and expect them to be remedied under the express terms of the contract relating to defects liability. This practice will rarely be challenged by the main contractor, as he will be contractually liable for the costs of remedying defects in his work which are due to his own default, and it will usually be cheaper for him to carry out the remedial work himself than to pay the costs of rectification by another contractor. However, in times of recession when property developers may be reluctant to take over buildings if they do not have a tenant or purchaser for them, the precise wording of the contract as to the contractor's duties during the defects liability period is sometimes used as a justification for delaying the issue of the practical completion certificate.

Within 14 days of the end of the defects liability period, the architect (or the employer under JCT 81) must send a schedule of any outstanding defects to the contractor, who must remedy them within a reasonable time of receipt of that schedule. When this has been done, the architect issues a certificate of completion of making good defects (described in JCT 81 as a notice of completion of making good defects), which is the trigger for the release of the second half of the contractor's retention.

JCT related sub-contracts

The provisions regarding defects liability in the sub-contracts are very similar to those in the main forms. However, the defects liability period is likely to be twelve months rather than six for specialist work, particularly mechanical and electrical services, to allow the employer to monitor the performance of the system in summer and winter conditions.

Practical completion of the sub-contract works may occur many months before practical completion of the main contract works, for example the structural steelwork and piling sub-contractors will have completed their work before most other trades have even started work on site. The JCT related sub-contracts deal with this by requiring the sub-contractor to accept a similar liability to that of the main contractor under the main contract to remedy defects in his work, and therefore the sub-contractor's contractual duty to remedy defects in his work will not expire until the end of the defects liability period applicable to the main contract works.

As mentioned earlier, one of the main incentives for both main contractors and sub-contractors to remedy defects in their work is the fact that the employer will be holding the second half of the retention fund during the defects liability period. However, nominated sub-contractors are entitled to early final payment on the expiry of twelve months after practical completion of their works, which will include the release of outstanding retention. Although they will continue to be liable to rectify defects in their work until the main contracts defects liability period has expired, they may be unwilling to do so. If this occurs, the main contract provides for another sub-contractor to be nominated to rectify the defects, although this will be at the main contractor's expense, because he is contractually liable for defects in work executed by nominated sub-contractors.

Works Contract/2 also contains provisions for early final payment to works contractors, but these are operated at the option of the employer or his architect, and are subject to a satisfactory indemnity being given to the management contractor by the works contractor in respect of latent defects.

Defects appearing after the expiry of the defects liability period

Once the relevant defects liability period has expired, the contractor or sub-contractor can no longer be required to rectify defects in his work. However, if the defect is due to his negligence or breach of contract, he will remain liable for the cost of any rectification work, and any consequential losses which it may cause, until the expiry of the relevant limitation period. Limitation periods are discussed in detail at the end of this chapter, but the basic rule is that the contractor remains responsible for any defects in his work for six years after practical completion if the contract is signed, and twelve years if it is executed under seal or as a deed.

Contractors often ask whether they are legally entitled to be given the opportunity to carry out any rectification work themselves after the defects liability period has expired. The injured party is under a duty to mitigate his loss (i.e. not to make it any worse than it needs to be), so as it will frequently be cheaper for the contractor to remedy his own work – both because he will be familiar with it and because he will not expect to make a profit on it as another contractor would – a failure to give the contractor such an opportunity is probably a breach of the duty to mitigate.

However, such a failure does not exonerate the contractor from his responsibilities in respect of the defective work, it simply means that the injured party will be unable to claim any more than it would have cost the contractor to rectify the work himself. It should also be borne in mind that the duty to mitigate will not require a building owner to invite the contractor to remedy defects in his work if it was so badly executed that it has totally destroyed the building owner's confidence in the abilities of that contractor.

Measure of damages for defective work

The cost of putting right a defect is often greatly exceeded by the costs which arise as a consequence of it, for example, damage to other parts of the building, and in serious cases, loss of use of the building which may entail loss of rent or loss of profits. Where a defect constitutes a breach of an express or implied term of the contract, the person who executed the defective work or supplied the defective goods or materials will be liable for damages.

The object of damages for breach of contract is to put the injured party in the same position he would have enjoyed if the contract had been properly performed. The leading case on damages for breach of contract is *Hadley* v. *Baxendale* [1854] 9 Ex. 341. That case laid down the rule that there are two classes of loss which can be recovered as damages for breach of contract:

(i)   all the loss which arises naturally, that is in the usual course of things, from the breach of contract; *and*

(ii) if there are special circumstances which increase that loss, the increased amount will be recoverable provided the contractor or subcontract was aware of those special circumstances before the contract was made.

The second type of loss might occur where a building was to be let for a rent much higher than one would normally expect for that type of property; only if the contractor or sub-contractor was aware of level of rent to be charged when he entered into his contract would he be liable to compensate the employer for its loss.

Limitation periods

A limitation period is the length of time during which legal action must be commenced in order to enforce a legal right. Once the relevant period has expired, if anyone attempts to sue for damages, it will be a complete defence for the defendant to plead that the limitation period has expired. This is sometimes described as pleading that the action is 'time-barred' or 'statute-barred'.

Legal liability must be finite for a number of reasons. Firstly, it is undesirable for the threat of legal proceedings to hang over the heads of potential defendants indefinitely. Potential plaintiffs must also be encouraged to start legal proceedings as soon as reasonably practicable after they have suffered loss, because the longer the delay in commencing proceedings, the more likely it is that documents will be lost and personal recollections will be unreliable, thus increasing the chance of an unjust result.

The action which must be taken within the relevant limitation period is the issue of a writ (if the proceedings are to be held in court) or the service of a notice of arbitration (if the dispute is to be resolved in arbitration). If there is a dispute as to which forum should be used, it is common practice both to issue a writ and to serve a notice of arbitration to ensure that both avenues remain open.

It is not sufficient simply actively to commence negotiations before the end of the limitation period, or to threaten proceedings or to instruct solicitors: the only thing which stops the limitation 'clock' is the formal commencement of proceedings by a writ or notice of arbitration.

The limitation periods applicable to sub-contractors will vary depending whether they are being sued for breach of contract or negligence.

In contract

The limitation period applicable to an action for breach of contract will be six years if the relevant contract was executed under hand (i.e. it was signed and the document was not a deed), or twelve years if it was executed under seal or as a deed.

The limitation period begins to run on the date the cause of action accrues, which is the date the breach of contract **occurs, not** the date it is discovered. Most commentators agree that, because the sub-contractor's basic obligation under the JCT related sub-contracts is to carry out and complete the sub-contract works, the date of breach will be the date of practical completion of the sub-contract works, notwithstanding the fact that the defective work may actually have been executed some months earlier.

Special rules apply to actions under indemnity clauses, where the cause of action will not arise until the loss against which the indemnity has been given is established. This can result in a quite significant extension of the limitation period normally applicable to a sub-contract.

For example, all of the JCT related sub-contracts contain a clause under which the sub-contractor indemnifies the main contractor against the consequences of any breach by him of the sub-contract. Assuming both main and sub-contracts are executed as deeds, the employer can make a claim against the main contractor for defective work by any sub-contractor until twelve years after practical completion of the main contract works. It is not until he does so that the cause of action under the sub-contract accrues, thus giving the main contractor a further twelve years to commence proceedings against the sub-contractor.

In tort

Limitation periods for actions to recover damages for negligence in respect of latent (i.e. hidden) damage are prescribed by the Latent Damage Act 1986. This is an immensely complex piece of legislation, in which Parliament strove to strike a balance between the rights of an injured party to redress and the need for finite legal liability. The Act has been widely criticised for failing to achieve either objective.

The Act provides that the limitation period for actions in negligence involving latent damage is whichever expires the later of:

six years from the date the damage occurred to the building (described in the Act as the date the cause of action accrued)

or

three years from the date the person who has the right to sue had the knowledge of the material facts relating to the damage or should have been aware of those facts

subject to

a fifteen year longstop, running from the date of the negligent act, after which no action may be taken.

In the context of the Act, 'knowledge' includes that which might reasonably have been acquired from observable or ascertainable facts, including those ascertainable by an appropriate expert, provided it was reasonable to expect the injured party to employ an expert.

These rules are best understood by applying them to a practical example. Suppose that, in 1980, a sub-contractor designed and installed a system for transporting chemicals in pipes laid in a concrete floor. The sub-contract works were practically complete in January 1981. The chemicals being carried by the pipes cause pinholing and corrosion, which expert evidence establishes began to occur in 1983. It was not until 1993 that the building owner noticed stains and cracks in the floor.

In this example, the damage began to occur in 1983, therefore the building owner had until at least 1989 to issue a writ. However, assuming that the damage should not have been discovered before it actually became apparent in 1993, the building owner will have additional time to issue a writ. This will not be the full three years from the date of discovery, because of the effect of the longstop. As the latest date the sub-contractor could have committed the negligent act is just before practical completion on 1 January 1981, in this example, the limitation period will expire on 1 January 1996, because of the operation of the 15 year longstop.

Fraud, concealment or mistake

The limitation periods summarised in this chapter will not apply where there has been fraud, deliberate concealment or mistake. Even the 15 year longstop under the Latent Damage Act is displaced: the injured party has six years from the date he did or could reasonably have discovered the fraud, concealment or mistake in which to issue a writ.

The extent to which the normal progress of building work which covers up defective work can constitute 'deliberate concealment' and therefore displace the limitation period has been considered by the courts on a number of occasions. In *Applegate* v. *Moss* [1971] 1 QB 406 the Court of Appeal said that 'if a builder does his work badly, so that it is likely to give trouble thereafter, and then covers up his bad work so that it is unlikely to be discovered for some years, then he cannot rely on the statute (i.e. the Limitation Act) as a bar to the claim.' However, it must be more than shoddy or incompetent work which is covered up in the normal course of building; there must be an element of deliberately turning a blind eye to work which any competent builder would realise should not be allowed to remain.

6 Sub-Contract Price and Variations

If there has ever been a project where the client didn't change his mind about what he wanted at some stage, no-one in the building industry is admitting it. In any event, when work starts on site the design is rarely complete, so variations are a fact of life on virtually all contracts.

As a result, the clauses which deal with them are among the most closely scrutinised at both main contract and sub-contract level. This chapter therefore sets out the definition of a variation adopted in the JCT forms, explains who has the power to require variations in a sub-contractor's work and summarises the rules for their valuation.

An understanding of the concepts of lump sum and remeasurement contracts and the function of a bill of quantities is also helpful when assessing what constitutes a variation, and so these issues have been addressed at the beginning of this chapter.

Lump sum and remeasurement contracts

A lump sum contract is an agreement to carry out a whole project or package of work for an agreed amount of money – for example to supply and install three lifts for £200 000. Regardless of the amount of work and materials required to supply those lifts, in principle the sub-contractor will only be entitled to be paid the sum he has quoted. Any rights to additional payment must arise under the express terms of the sub-contract, for example if there are variations or an applicable fluctuations clause.

In contrast a remeasurement or 'measure and value' contract is one where the sub-contractor submits a schedule of rates or prices which will be used to value the amount of work which he actually carries out. The sub-contractor's tender sum is simply an estimate based on the amount of work which it is expected will be required of him, and serves no purpose other than acting as a basis for selection.

Most of the standard forms of sub-contract with which this book is concerned are drafted in such a way that they can be used on either lump sum or remeasurement contracts – the exceptions are the two sub-contracts for use with the Intermediate Form: NAM/SC and IN/SC, which have no remeasurement option.

The inclusion of the two options can make forms such as NSC/C and DOM/1 appear more complicated than they are, as they generally contain two sets of very similar valuation provisions – one for valuing variations where the sub-contract is lump sum, and one for valuing all of the works, including variations, where it is subject to remeasurement.

Taking DOM/1 as an example, the two alternatives are dealt with as follows. The articles of agreement will indicate the basis on which the sub-contractor was invited to quote. Where the sub-contract is on a lump sum basis Article 2.1 will apply, which describes the sum quoted by the sub-contractor as the 'Sub-Contract Sum'. Clause 15.1 of DOM/1 provides that sum will be the price which the sub-contractor is entitled to be paid for his work, subject only to such adjustments as are expressly permitted by the sub-contract in respect of variations, etc.

If it is a remeasurement sub-contract Article 2.2 will apply, which describes the sum quoted as the 'Tender Sum'. Clause 15.2 states that the sub-contract works will be 'subject to a total remeasurement' in accordance with the terms of the sub-contract. This means that all of the work executed by the sub-contractor will be valued in accordance with what are described as 'the valuation rules' which, as explained earlier, are drafted in virtually the same terms as the rules for valuing variations on a lump sum contract.

Strictly speaking, the sub-contract should only be let on a lump sum basis where the work which the sub-contractor will be required to carry out is reasonably well defined at tender stage, either by a bill of quantities or by drawings and a specification. If the work cannot be properly defined, for example because the design is incomplete, a remeasurement contract is preferable because it will entitle the sub-contractor to be paid for all the work he executes at the rates he has quoted. If a lump sum sub-contract was used in such circumstances, it would carry a very significant element of risk for the sub-contractor, as he would only be entitled to extra payment if he could demonstrate that any further design details he was given constituted a variation, which will often be a difficult task.

Bills of quantities

Bills of quantities describe and quantify the work to be carried out under the contract. They are usually prepared by the client's quantity surveyor in accordance with a standard method of measurement. All of the sub-contracts with which this book is concerned stipulate that where bills of quantities are a sub-contract document, they should be prepared in accordance with the *Standard Method of Measurement for Building Works, 7th Edition* (*SMM7*) published by the RICS and the BEC.

There is, however, some scope for deviations from the *SMM*, as all of the sub-contracts state that specified items may be treated in a different manner to that required in the *SMM* provided the bills themselves make this clear.

If the bills contain errors in description or quantity, or omit items, or depart from the *SMM* other than where the bills themselves specify that this has been done deliberately, the error must be corrected by an architect's instruction or contractor's direction. This will be treated as if it were a variation, and will thus entitle the sub-contractor to additional payment and an extension of time and loss and expense where appropriate.

Where bills of quantities are a sub-contract document, all of the sub-contracts state that the quality and quantity of the work which is included in the sub-contractor's sub-contract sum or tender sum is deemed to be that which is set out in the bills of quantities. Anything which the sub-contractor is instructed to supply which is additional or different to what is set out in the bills will therefore entitle him to extra payment.

This principle mitigates, to some extent, the risk element of a lump sum contract for the sub-contractor, because it will be clear that anything which is not in the bills will be a variation. The sub-contractor is much more at risk where bills of quantities are not a sub-contract document, and he is required to supply everything shown upon and described by a specification and drawings.

Prime cost and provisional sums

The terms 'prime cost sum' and 'provisional sum' are often treated as interchangeable but in fact they each have a precise and different meaning, particularly if bills of quantities are part of the contract documents.

A prime cost sum is a sum included in the price for the main contract works to be expended on work, materials and goods to be supplied by a nominated sub-contractor or a nominated supplier. When he tenders for the main contract, the contractor adds an amount to the prime cost sum to allow for his profit and any attendances he is required to provide. If tenders have not been obtained from sub-contractors and suppliers when the main contractor is invited to tender, an indicative sum is inserted in the bills or other priced document as a prime cost sum, and when the exact sum is known the contractor's allowance for profit and attendances is adjusted accordingly.

A provisional sum is a sum of money inserted in the bills or other priced document to cover the cost of anything which cannot be detailed accurately at tender stage or which the employer may, but not necessarily will, require. When the contractor is instructed to expend the provisional sum, the sum is deleted from the priced document and the work actually carried out will be measured in accordance with the valuation rules in the contract.

Where bills of quantities prepared in accordance with *SMM7* are a contract document, there are two types of provisional sum which may be used: a provisional sum for defined work, and a provisional sum for undefined work. General Rule 10 of *SMM7* states that a provisional sum for defined work can only be included in the bills when certain information about that work can be given to the contractor, namely:

- the nature and construction of the work;
- a statement of how and where the work is fixed to the building and what other work is to be fixed to it;
- a quantity or quantities indicating its scope and extent;
- any specific limitations on method, sequence or timing of the work identified in accordance with Section A35 of *SMM7*.

If that information cannot be given, the provisional sum will be for undefined work. The distinction between the two is not just a semantic one, as the contractor is deemed to have allowed for programming, planning and pricing preliminaries for work which is the subject of a provisional sum for defined work, and so will not be entitled to recover additional preliminaries costs or to an extension of time when it is expended.

The rules for valuing work carried out pursuant to an instruction to expend provisional sums are discussed in more detail under the heading of valuation.

Definition of a variation

All of the JCT main contracts and their related sub-contracts define variations in a similar way, and so no distinction has been drawn between the different forms of sub-contract in this section.

The first principle to bear in mind is that, in order to constitute a variation, a change must be required by either an instruction of the architect or a direction of the main contractor, depending on the form of sub-contract which applies (see later for a discussion as to who can instruct variations under the various forms).

Secondly, many people do not appreciate that there are two quite different types of variation:

- a variation in the work itself; *and*
- a variation in the circumstances under which the work is executed.

Everyone recognises that adding an extra floor to the building is likely to be a variation, but an instruction not to work before 9 a.m. is likely to be treated as a 'claim' item, rather than as a variation. Given that 'claims' are almost invariably reduced by negotiation, and variations tend to be valued in accordance with the contract, this could be an expensive misconception for a sub-contractor.

Where the sub-contract is on a lump sum basis, a variation in the work itself is defined as an alteration or modification in the design, quality or quantity of the sub-contract works as defined in the numbered documents (i.e. the bills, specification, etc. on which the sub-contractor tendered) including:

- the addition, omission or substitution of any work;
- the alteration of the kind or standard of any of the materials or goods to be used in the sub-contract works;
- the removal from the site of any work, materials or goods, provided they are not defective.

The definition of a variation in the work on a remeasurement contract is the same, save that there is no reference to variations in the quantity of work required. Changes in quantity do not need to be included in the definition on remeasurement contracts because, as explained above, any quantities given at tender stage will be indicative only, and all of the work will automatically be remeasured.

Variations in the circumstances under which the work is executed are defined as a change in any obligations or restrictions imposed in the sub-contract documents or any new obligations or restrictions in regard to:

- access to the site or use of any specific parts of the site;
- limitations of working space;
- limitations of working hours;
- the execution or completion of the work in any specific order.

Thus if the sub-contractor had been warned in the sub-contract documents that the site would close at 2 p.m. on Fridays he must take that into account in his tender, but if that restriction is introduced during the currency of the contract, it will constitute a variation.

Power to instruct variations

Sub-contractors tend to have particular problems in identifying the person who is contractually entitled to issue them with variation instructions, because it is often not the person who, in practice, controls and supervises their work. For example, a mechanical services sub-contractor is likely to have a much closer working relationship with the consulting engineer than with either the architect or main contractor, but it is highly unlikely that the engineer will have the contractual power to vary the sub-contractor's work.

Each of the standard sub-contracts has slightly different rules regarding who can instruct variations and the procedure for doing so, so the position under each document is summarised below.

Nominated sub-contract NSC/C

Who may instruct variations

Only the architect may vary the nominated sub-contract works. The main contractor has the power to issue 'any reasonable direction' regarding the sub-contract works, but as the definition of a variation is limited to changes which are required by an instruction of the architect, it is reasonably clear that this power does not extend to requiring variations.

The main contractor is expressly obliged by clause 3.3.1 of NSC/C to pass on to the nominated sub-contractor all architect's instructions (including those which require a variation) 'forthwith' — in other words immediately.

Oral instructions

In principle, all instructions and directions should be issued in writing but, for a variety of reasons, they are very often given orally. NSC/C contains some quite detailed rules governing the validity of oral instructions, and although the procedures may look cumbersome they are extremely important for the simple reason that the sub-contractor will only have a clear right to additional payment if a variation is ordered in a properly issued instruction.

The procedure which should be followed when an oral instruction is issued to the sub-contractor is laid down in clause 3.3.3 of NSC/C, and may be summarised as follows:

- the sub-contractor is not under an immediate obligation to comply with any instruction or direction not issued in writing;
- the sub-contractor must confirm any oral instruction in writing to the main contractor within 7 days of receiving it;
- if the contractor does not dissent from the sub-contractor's written confirmation within 7 days from receipt, the instruction takes effect on the expiry of that 7 day period;
- if the contractor himself confirms the oral instruction or direction in writing within 7 days of its issue, it takes effect immediately as a properly issued instruction, and the sub-contractor is not obliged to issue a written confirmation himself.

Sub-contractors often ignore these rules and begin to comply with oral instructions either as soon as they receive them or when they have issued their own written confirmation. Although there may be sound practical reasons for doing this, at that stage the instruction has no contractual effect and so the sub-contractor will be working at his own risk. Any work which the sub-contractor carries out will not entitle him to extra payment under the sub-contract if the main contractor does dissent from the sub-contractor's written confirmation, or if the sub-contractor himself fails to confirm the instruction in writing.

Although clause 3.3.3.2 of NSC/C states that the main contractor may confirm an oral instruction in writing after the relevant work has been carried out, provided he does so before final payment is made to the sub-contractor, that clause creates a right rather than an obligation on the main contractor. The only circumstances where the nominated sub-contractor can force the main contractor to issue a retrospective confirmation is if the architect has chosen to confirm the relevant oral instruction under the main contract.

Right to object to variations

NSC/C gives the nominated sub-contractor an express right to object to variations in very limited circumstances, namely where:

- the variation is in respect of obligations or restrictions relating to the way in which the work is carried out; *and*
- the sub-contractor has made a reasonable objection in writing.

The sub-contractor has no express right to object to variations in the work itself, and therefore from the face of the contract it would appear that even if the amount of work required is dramatically increased or reduced, the sub-contractor would continue to be bound by the sub-contract and the rates and prices he had originally quoted.

There may be an implied right to refuse to carry out variations under the sub-contract if they fundamentally alter the nature of what the sub-contractor has contracted to supply, but a recent case in the Court of Appeal shows how difficult this will be to prove.

In *McAlpine Humberoak Ltd* v. *McDermott International Inc.* (1992) 82 BLR 1 McAlpine had contracted to supply four pallets which were to form part of an oil rig in the North Sea. A very considerable number of drawings was issued to McAlpine in the early stages of their work, and eventually two of the four pallets were omitted from their contract.

Although McAlpine had neither pleaded nor argued that the effect of the variations was to frustrate the contract, this was the finding of the judge at first instance.

Where frustration occurs, the contract is brought to an end and the sub-contractor has a statutory right under the Law Reform (Frustrated Contracts) Act 1943 to be paid for any 'valuable benefit' which the main contractor has received, which is assessed by the court as being just in the circumstances of the case. The figure produced by this exercise is usually similar to *quantum meruit*, and will reflect what the work is worth.

In the *McAlpine* case, the judge at first instance awarded damages to the contractor based on *quantum meruit*, which he calculated as their costs plus a 10 per cent allowance for profit.

The Court of Appeal, however, took a very different view. They found that the variations had neither transformed the contract into a different contract nor distorted its substance and identity. It had remained a contract for the construction of pallets, and therefore its terms continued to bind the parties.

This decision will make it much more difficult for contractors and sub-contractors to argue that they can refuse to carry out substantial variations or that they invalidate the contract and must therefore be paid for on a *quantum meruit* basis. Perhaps if the sub-contractor had contracted to put a roof on a hospital and by variation this was omitted and he was required to put a roof on a private house he might win this particular argument, but it is now clear that variations in the amount of work required, however significant, will rarely amount to a change in scope.

Domestic sub-contract DOM/1

Who may instruct variations

Under DOM/1, only the main contractor has the power to vary the sub-contract works. Under clause 4.2.1 he may issue any reasonable direction to the sub-contractor, including directions which require a variation.

However, the sub-contract acknowledges that the source of many of the instructions to the domestic sub-contractor will be the architect, by providing that any written architect's instruction issued under the main contract will be deemed to be a direction of the main contractor **provided** it is issued by the contractor to the sub-contractor. An instruction issued directly by the architect to the sub-contractor would not satisfy this requirement, and although the sub-contractor is unlikely, in practice, to have any difficulty in securing payment for it, he should remember that it will not have been validly issued under DOM/1 if it has not been passed through the main contractor.

Oral instructions

The rules in DOM/1 relating to the validity of oral instructions are exactly the same as those in NSC/C, save that where the main contractor has the option retrospectively to confirm an oral instruction with which the sub-contractor has complied, it will be entirely at his discretion, regardless of whether the architect has retrospectively confirmed the relevant instruction under the main contract.

Right to object to variations

A domestic sub-contractor's rights to object to a variation under DOM/1 are exactly the same as those of a nominated sub-contractor under NSC/C.

Domestic sub-contract DOM/2

Who may instruct variations

The rules in DOM/2 in this regard are exactly the same as those in DOM/1.

Oral instructions

The rules in DOM/2 are the same as those in DOM/1.

Right to object to variations

Because a domestic sub-contractor working under DOM/2 is likely to be involved in design, he has additional rights to refuse to comply with a variation instruction to those in DOM/1.

Under clause 4.2.1 of DOM/2, the main contractor is not entitled to order a variation which results in an alteration or modification in the design of the sub-contract works without the sub-contractor's consent, which must not be unreasonably delayed or withheld. This is a more powerful right than a right simply to object, as the sub-contractor's consent is a pre-condition to any variation being instructed.

Named sub-contract NAM/SC

Who may instruct variations

NAM/SC provides that only the main contractor may issue instructions requiring variations in the named sub-contract works. It does, however, contain provisions identical to those in DOM/1 regarding architect's instructions issued to the named sub-contractor being deemed to be directions of the contractor.

Oral instructions

Neither the Intermediate Form nor its sub-contracts provide for the issue of oral instructions. They simply provide that all architect's instructions and contractor's directions must be issued in writing.

Although this is undoubtedly a laudable principle, it creates enormous problems, because oral instructions will inevitably be issued in practice. This means that the sub-contractor will be confronted with an unattractive choice: comply and risk not being paid, or refuse and risk antagonising the architect and main contractor. Although he may face a similar dilemma under the other forms, at least there he has the option of confirming the instruction or direction in writing himself, but under NAM/SC the sub-contractor's written confirmation will have no contractual effect whatsoever.

One practical solution is to ensure that all senior site staff are armed with duplicate pads, and whenever they are given a direction by the main contractor, they simply ask him to write it down, give him a copy and retain one for their records. It is worth remembering that any form of writing is sufficient to make the direction valid under the contract – there is no need for an official instruction form, or even for it to be on headed paper provided it is signed by an appropriate representative of the main contractor.

Right to object to variations

The named sub-contractor's rights are exactly the same as those of a domestic sub-contractor under DOM/1.

Domestic sub-contract IN/SC

IN/SC is identical to NAM/SC in all respects relating to the issue of directions requiring a variation.

Works Contract/2

Who may instruct variations

Works Contract/2 is different to all of the forms discussed above in that it formally defines both an 'Instruction' and a 'Direction'.

Instructions are architect's instructions issued under the Management Contract and passed on to the works contractor by the management contractor, whilst directions are the management contractor's own reasonable requirements.

Both instructions and directions will amount to variations if they fall within the definition of a variation set out at the beginning of this section, so it is clear that the management contractor, as well as the architect, has the power to vary the works contract works.

Oral instructions

If the management contractor issues any instruction or direction to the works contractor which is not in writing, a much simpler procedure applies than under other standard forms which permit oral instructions.

Either the management contractor or the works contractor may confirm the oral instruction within seven days of its issue. If neither does so, it simply has no effect under the contract. There is no contractual mechanism for the management contractor to dissent from the works contractor's written confirmation if he disagrees with it or if he has changed his mind, but presumably in such circumstances he would simply issue a further direction clarifying his intention.

The works contractor is in a much stronger position than a sub-contractor under any of the other forms when an oral instruction is issued, because his own written confirmation converts it into a validly issued instruction without the need to wait for any dissent. He should therefore be entitled to be paid for any work he does pursuant to his written confirmation, even if the management contractor does subsequently dissent from it.

Right to object to variations

A works contractor's right to object to variations is the same as a nominated sub-contractor's, save that because his work may include design, he also has express rights regarding any instruction or direction which involves a modification to any part of the works which he is designing. On the issue of such an instruction or direction, the works contractor must tell the management contractor whether he consents to comply with it (note that, unlike DOM/2, his consent is not a pre-condition to the issue of the instruction). If he does object or withhold his consent, the management contractor must refer the matter to the architect for his decision.

Although the contract is silent on the point, if the works contractor is instructed by the architect to comply with the instruction despite his objection, it is likely that the works contractor would not be liable if that aspect of the design proved to be defective.

Valuation of variations and provisional sums

All of the JCT standard forms and their related sub-contracts which are considered in this book contain broadly similar provisions for the valuation of variations and provisional sums.

The basic valuation rules which are common to all of the forms are set out below, followed by brief notes on each form of sub-contract summarising its particular characteristics.

The standard valuation rules

There are four types of work which fall to be valued under the valuation rules in a lump sum sub-contract:

(i) *additional or substituted work which can be measured*

The natural reaction of most quantity surveyors when valuing extra work is to use the rates and prices in the bills or other priced document. However, under the contract these will only be applicable where:

- the work is of a similar character to that set out in the contract documents; *and*
- the work is executed under similar conditions; *and*
- there is no significant change in quantity.

If the varied work is of a similar character to the work in the bills or other priced document, but there is a significant change in quantity or it is executed under different conditions, the bill rates will still be used as a basis for valuation, but a fair allowance should be made for the difference in conditions or quantity.

If the work is not of a similar character to that required by the contract documents, it should be valued at fair rates and prices. If the sub-contractor has heavily discounted his bill rates there is a strong argument for applying a similar discount to the calculation of those fair rates and prices, so the sub-contractor is unlikely to be able to convert a sub-economic tender into a profitable job by recovering payment for such work at cost plus a realistic allowance for overheads and profit. In the case of *Laserbore Ltd* v. *Morrison Biggs Wall Ltd* (1993) Building Law Monthly Vol. 10 Issue 11 it was held that a contractual right to 'fair and reasonable payment' meant a fair commercial rate for the services provided rather than payment on a costs plus basis.

(ii) *work for which an approximate quantity has been included in the bills*

SMM7 is the first standard method of measurement to allow for the use of approximate quantities in an otherwise firm bill. According to General Rule 10, they should be used where work can be described in accordance with the requirements of the *SMM*, but the quantity of work required cannot be accurately determined.

If the approximate quantity is a reasonably accurate forecast of the amount of work actually required, the rate in the bills will be used to value it. If it is not a reasonably accurate forecast, the rate is still used as a basis for the valuation, but a fair allowance is made for the difference in quantity.

If the approximate quantities work has been varied in some other way than in quantity – in other words rather than requiring more or less of the work than was indicated in the bills, the sub-contractor is instructed to provide something different, or the work is executed under different conditions, that change will be valued under the normal rules for valuing variations set out under (i) above.

(iii) *additional or substituted work which cannot be measured-daywork*

If additional or substituted work cannot be valued by measurement, it will be paid for on the basis of daywork. There are two ways in which daywork rates can be established: either the sub-contractor can quote a schedule of daywork prices, or the daywork rates can be calculated by applying a percentage addition to the prime cost of the labour, materials, and plant used to carry out the work.

The sums which should be included in the prime cost are laid down in a number of standard definitions agreed between the RICS and certain contractors' trade associations.

BEC has agreed a definition with the RICS which covers all general building work, and this is the most widely used definition of prime cost. In addition, the Heating and Ventilating Contractors' Association (HVCA), the Electrical Contractors' Association (ECA) and the ECA of Scotland have also agreed their own definitions which refer to rates of pay agreed under the wage-fixing body which represents their specialist sector. The BEC definition refers to the wages set in accordance with the NJCBI National Working Rule Agreement.

The different definitions of prime cost can be a source of conflict between main contractors and sub-contractors, as the sub-contractor will naturally wish to use the relevant specialist definition because it will most closely reflect the rates he is paying his employees. The standard forms generally allow for this by providing in the main contract that where the work is within the province of a specialist trade which has agreed a definition of prime cost with the RICS, its prime cost should be calculated in accordance with that definition. Where a specialist definition is applicable to the sub-contract works, it should be specified in the tender document (in sub-contracts such as NSC/C and Works Contract/2) or in the appendix to the sub-contract (in sub-contracts such as DOM/1).

Despite these manifestly sensible arrangements there are some main contractors who insist on paying sub-contractors for daywork on the basis of the RICS/BEC definition. Unless appropriate amendments have been made to the standard form, this will be a breach of the sub-contract.

If there is no standard definition of prime cost applicable to the particular sub-contract trade and their labour is not covered by the NJCBI agreement, the most sensible course is for the sub-contractor to include a schedule of daywork rates in his tender.

Many sub-contractors regard daywork as an opportunity to make money, but in practice it can be the source of endless disputes, particularly as to whether the hours recorded on the sub-contractor's daywork sheet are an accurate record of how long the work actually took. All of the standard sub-contracts require the sub-contractor to deliver what they describe as 'vouchers' (i.e. daywork sheets) to the contractor for verification of the hours worked and the plant and materials used at the end of the week following that in which the work was executed.

This is usually hopelessly late: a much more efficient (although admittedly time-consuming) approach is to ensure that a representative of the contractor is present when the daywork is executed and that the daywork sheet is signed the same day. Although this should avoid disputes as to the time worked, the sub-contractor should remember that a signed daywork sheet is not an admission by the contractor that daywork is the proper method to value the variation, or indeed that the work is a variation at all.

(iv) *omissions*

Omissions are valued at the rates and prices set out in the bills or other priced document. If work is omitted and given to another sub-contractor, as opposed to not being carried out at all, it is generally thought that this constitutes a breach of contract and that the sub-contractor can also recover loss of profit on the omitted work. Strictly speaking, the claim for loss of profit should be framed as a claim for breach of contract rather than recovered as part of the valuation of the variation.

In addition to the sums due under these valuation rules, the sub-contractor will also be entitled to recover:

- additional preliminaries, provided the work is neither the subject of a provisional sum for defined work nor valued on the basis of daywork;
- the additional costs of carrying out unvaried work due to a variation changing the conditions under which that work is executed;
- a fair valuation of any variation which does not relate to additional, substituted or omitted work (e.g. a variation in the access to the site stipulated in the contract documents);
- a fair valuation of any work or liabilities directly associated with a variation which it is not reasonable to value in accordance with the valuation rules.

If variations disrupt the regular progress of the sub-contract works, the sub-contractor will also be entitled to loss and expense, but that is paid under the loss and expense provisions of the contract and forms no part of the valuation of the variation.

An alternative method of dealing with variations is to pre-agree a price for the varied work before it is executed. This has the great merit of speed and simplicity, although it does involve an element of risk for both contractors and sub-contractors as if the variation actually costs much more than they anticipate they will not usually be entitled to recover the additional costs. The only standard form which provides a detailed mechanism for such preagreement is the Design and Build Form 1981, but it has not been stepped down into sub-contract DOM/2. It is understood that the JCT is working on similar provisions for JCT 80, but at the time of writing these have not been published.

Particular characteristics of the different sub-contract valuation rules

Nominated sub-contract NSC/C

As nominated sub-contract work is the subject of a prime cost sum, the quantity surveyor values variations in that work.

Domestic sub-contract DOM/1

DOM/1 does not expressly provide for who is responsible for valuing variations in the domestic sub-contract work — the form simply provides for the sub-contractor to be paid the value of variations which is established in accordance with the basic valuation rules. In practice, the main contractor will pay the sub-contractor what he considers to be the proper value of varied work: if the sub-contractor disputes it his only remedy will be arbitration.

Domestic sub-contract DOM/2

The rules for valuing variations are identical to those in DOM/1.

Named sub-contract NAM/SC

The valuation rules in the Intermediate Form and its associated sub-contracts generally follow the basic rules already described, but are in an abbreviated form. There are, however, two aspects of the rules which are unique to IFC 84 and its sub-contracts, namely:

- daywork can be used wherever it is an appropriate basis for a fair valuation, not just where it cannot be measured;
- the contracts expressly provide for the value of the varied work to be agreed before the work is executed, and only if that cannot be done will the valuation rules operate. It does not, however, provide a mechanism for such an agreement to be reached.

As under DOM/1, the contract does not stipulate who should value variations, but in practice it is done by the main contractor.

Domestic sub-contract IN/SC

See the comments on NAM/SC.

Works Contract/2

Works Contract/2 provides for the value of variations to be included in architect's certificates; therefore they are valued by the quantity surveyor.

7 Payment and Set-off

In 1974, in a case called *Gilbert-Ash (Northern) Ltd* v. *Modern Engineering (Bristol) Ltd*, Lord Denning said:

'There must be a "cash flow" in the building trade. It is the very lifeblood of the enterprise.'

His words have been repeated so often they have become a cliche, but like most cliches, they are absolutely true. If money does not cascade down the contractual chain the building process will eventually come to a halt. The contractual rules governing payment to contractors and sub-contractors usually acknowledge this, by providing for regular interim payments and for powerful remedies if they are unjustifiably withheld.

Sub-contractors are in an especially vulnerable position when it comes to payment. They are often smaller concerns than the main contractor, and therefore more dependent on receiving their money on time, particularly on large contracts which represent a significant proportion of their total turnover. Main contractors are generally better organised, and so are able to manipulate the contractual provisions to their own advantage. This has turned the payment and set-off provisions in the standard forms of sub-contract into something of a battleground, and many of the building disputes which have come before the courts in recent years relate to the payment of sub-contractors.

This chapter describes how the standard forms of sub-contract deal with payment and covers the closely related subject of the main contractor's right to make a set-off from those payments. As many smaller sub-contracts are let on the basis of a simple order containing no detailed terms and conditions, it also explains the payment terms the law will imply into the sub-contract where there is no express provision for interim payment.

Interim payment

Nominated sub-contract NSC/C

Interim payments to nominated sub-contractors are based on architect's certificates issued under the main contract. Clause 35.13 of JCT 80 provides that the architect must calculate the sums due to nominated sub-contractors in accordance with the relevant provisions of NSC/C, and must then direct the main contractor as to how much of the sum certified in each interim certificate is due to each nominated sub-contractor.

111

The architect is generally obliged to issue interim certificates under the main contract at monthly intervals (unless a different period is stated in the appendix) and the nominated sub-contractor is entitled to be paid the sum certified in his favour within seventeen days of the date each certificate is issued.

Applications for payment

NSC/C does not require the nominated sub-contractor to make any application for payment, although in practice this almost invariably occurs. The sub-contractor is able to ensure that his views on the sums due to him are passed on to the architect, as under clause 4.15.1 of NSC/C the contractor must, if so requested by the sub-contractor, apply to the architect to certify sums due to the nominated sub-contractor, and must include in that application any written representations which the sub-contractor wishes the architect to consider.

Main contractors often attempt to impose an obligation on the nominated sub-contractor to apply for payment at a particular time, and may even refuse to pay if he fails to do so. It is, however, unlikely that such a failure amounts to a breach by the sub-contractor. This is because the requirement to apply for payment is usually included in the numbered documents, and clause 1.6 of NSC/C makes it clear that if there is a conflict between the conditions of NSC/C and any of the other sub-contract documents, NSC/C will prevail. As NSC/C does not require the sub-contractor to apply for payment, the introduction of such a requirement in another document would probably have the effect of creating a conflict, thus allowing the sub-contractor to rely upon clause 1.6.

Even if the main contractor has physically altered the conditions of NSC/C to require applications for payment, if the sub-contractor is in breach the contractor could only withhold payment of any certified sums if the amendment was drafted in such a way as to make it clear that a proper and timely application was a condition precedent to payment.

Amounts due

The amounts to be included in interim certificates in favour of nominated sub-contractors are set out in clause 4.17.1 of NSC/C. The total sum is described as 'the gross valuation' and should consist of:

- the total value of work properly executed by the sub-contractor including variations, plus formula fluctuations, if applicable;
- the total value of goods and materials on site, provided they have not been prematurely delivered and are adequately protected against weather and other risks (note that there is **no** requirement to demonstrate a good title to the goods and materials once they are on site, provided they have been properly and not prematurely delivered);

- at the discretion of the architect, goods and materials off site (the circumstances in which the architect may exercise his discretion are covered later in this chapter);
- statutory fees and charges;
- loss and expense due to delay and disruption;
- tax or conventional fluctuations, if either are applicable.

The first three sums are subject to the deduction of retention at either 3 or 5 per cent before practical completion and at half that rate until the completion of making good defects. As the remainder are costs which the sub-contractor is entitled to recover in full, they are not subject to retention and one thirty-ninth is added to them to compensate for the two and a half per cent cash discount which the main contractor can deduct if he pays on time.

As the sub-contract work is valued on a cumulative basis, previous payments must be deducted from the gross valuation. It is important that the various deductions which are to be made are done in the right order, which is as follows:

(i)   retention
(ii) sums previously certified
(iii) cash discount

It is quite common for main contractors to deduct cash discount before deducting sums previously certified, which means that it is deducted over and over again from the sums due for the earlier work. This is a breach of NSC/C, which makes it clear that cash discount should be calculated on the basis of the net sum due, and should therefore be the last deduction to be made.

Sub-contractors perceive the architect's involvement in payment as one of the major benefits of nomination, but it is not a panacea. If the architect or quantity surveyor has undervalued the sub-contractor's work his only remedy is to arbitrate directly against the employer by using the 'name borrowing' provisions in clause 4.20 of NSC/C, and this can be a cumbersome procedure (for further discussion see Chapter 12 on Methods of Dispute Resolution).

Retention

In accordance with the normal practice in building contracts, retention will be deducted from payments due to the nominated sub-contractor so that the employer has a sum of money available to him in the event that the nominated sub-contractor is in breach, for example if he fails to remedy defects in his work.

NSC/C provides that the amount of retention to be deducted will be 5 per cent of those sums which are subject to retention, unless a lower rate is agreed and inserted in the main contract appendix. A footnote to JCT 80 recommends that on contracts worth over £500 000 the retention percentage should be not more than 3 per cent, but this is usually ignored and 5 per cent retention is the norm even on very large contracts.

As mentioned above, the retention percentage applied to interim payments due to the nominated sub-contractor reduces by half on practical completion of the nominated sub-contract works, thus releasing half of the retention fund. Note that the nominated sub-contractor does not have to wait until practical completion of the main contract works to receive the first half of his retention – a key point for early trades such as structural steelwork and piling.

The second half of the retention fund should be released on the issue of the certificate of completion of making good defects under the main contract, which may be some considerable time after the defects liability period under the sub-contract has expired. However, the sub-contractor may be able to secure release of his retention at an earlier stage under the early final payment provisions in clause 35.17 of JCT 80; these are discussed later in this chapter.

Retention money deducted from payments due to nominated sub-contractors has trust status in the hands of both the employer and the main contractor. This is an important protection for the sub-contractor, as it means that if the employer or main contractor becomes insolvent while they are holding the sub-contractor's retention, the sub-contractor should be able, at least in theory, to recover it in full. But in practice, the sub-contractor will find this right is almost impossible to enforce unless his retention money has been placed in a separate bank account before the insolvency.

This is because the protection afforded by trust status will be lost if the sub-contractor cannot identify his retention because it has become mixed with the other assets of the main contractor or employer. The sub-contractor must be able to 'trace' his own retention fund, that is to identify the money being held on his behalf, which is particularly difficult if it has been paid into an overdrawn bank account.

While the retention is being held by the employer, clause 30.5.3 of the private editions of JCT 80 allows the sub-contractor to protect himself by providing that the employer must, at the request of the main contractor or any nominated sub-contractor, place the retention in a separate bank account. (The local authorities versions of JCT 80 do not contain that provision on the grounds that a local authority cannot become insolvent.) A wise sub-contractor will avail himself of this protection if there is any doubt whatsoever as to the solvency of the employer, as if the money is not in a separate account, the mere fact that it has trust status will not allow him to recover it in full as the trust will not be fully constituted.

The effect of giving retention money trust status was considered by the courts in the case of *Wates Construction (London) Ltd* v. *Franthom Property Ltd* (1991) 53 BLR 23 where the employer had deleted the provision permitting the contractor to require him to pay the retention into a separate account. The court held that, notwithstanding that deletion, the first duty of the employer as trustee was to safeguard the retention fund, and so it would be a breach of trust to use the money for the purposes of his own business. He was therefore obliged to pay the retention into a separate account by the general law, and could not rely on the deletion of the express obligation to do so.

Employers rarely act in accordance with the principle established by *Wates* v. *Franthom* and will place the retention money in a separate account only when asked to do so. If they refuse, the sub-contractor can enforce his rights by applying to the courts for an injunction.

For most of the time, the sub-contractor's retention will be held by the employer rather than the main contractor, but there will be at least a few days when it is in the hands of the main contractor. Following *Wates* v. *Franthom* the main contractor should, in theory, place the money in a separate account during that time, but because of the short periods involved this is rarely done. Clause 4.22 of NSC/C places an express obligation on the main contractor to pay the money in a separate account if he tries to mortgage or charge his interest in the retention or if he fails to pay it on time.

If the main contractor becomes insolvent before the nominated sub-contractor's retention has been released by the employer, the sub-contractor should be able to recover his retention in full directly from the employer. Although direct payment is prohibited both by the general law and by the express terms of JCT 80 in the event of the main contractor's bankruptcy or liquidation, this will not apply to retention because of its trust status; this principle was confirmed in the case of *Re Arthur Sanders Ltd* (1981) 17 BLR 125.

A question which is often raised in relation to nominated sub-contractor's retention is whether, in the event of a default by the main contractor which is unrelated to the performance of the nominated sub-contractor, the employer can deduct money not only from the main contractor's retention, but from any retention held on behalf of the 'innocent' nominated sub-contractor.

The answer to this seems fairly clear, because under clause 30.1.1.2 of JCT 80 the employer may exercise any right of deduction against 'Retention' which is defined as including both the nominated sub-contractor's and the main contractor's retention. The nominated sub-contractor is therefore at risk in the event of a major default by the main contractor, although some commentators have suggested that the employer should exhaust the retention held on behalf of the main contractor before turning to the nominated sub-contractor's retention.

Cash discount

Clause 4.16.1.1 of NSC/C makes it absolutely clear that the main contractor's entitlement to deduct a two and a half per cent cash discount from sums due to nominated sub-contractors only arises if he pays the sub-contractor in accordance with the sub-contract – that is within seventeen days of certification under the main contract.

This restriction is often ignored by main contractors, who automatically deduct the discount no matter how late the payment. The fact that this is a straightforward breach of contract which could be challenged by the sub-contractor with relative ease is often obscured by references in the contract documents to the discount as 'main contractor's discount', or by claims by the main contractor that it is a payment to him for co-ordination and attendances. Under the standard forms of sub-contract neither of these contentions is correct: it is a discount for prompt payment and prompt payment alone.

The sub-contractor is not obliged to challenge the incorrect deduction of discount each month but could, if he wished to do so, reserve his claim that it had been wrongfully deducted until negotiation of the final account, when it can be a useful weapon in his armoury.

Payment for goods and materials off site

This is an important subject for many sub-contractors, particularly those such as structural steelwork and lift contractors who manufacture or fabricate the products they supply, and who therefore incur significant costs before starting work on site.

Nominated sub-contractors are entitled to be paid for goods and materials off site at the architect's discretion. This is not an unfettered discretion, in that the architect may not exercise it unless certain criteria, which are listed in clause 30.3 of JCT 80, have been satisfied. These requirements, which are aimed at ensuring that the employer will receive a good title to the goods and materials for which he is paying, may be summarised as follows:

- the goods and materials must be intended for incorporation into the works;
- there must be nothing else to be done to them before they are incorporated into the works (which means that unfabricated steel, for example, or a partly manufactured panel of cladding cannot be included in a certificate under these rules);
- they must have been set apart and clearly marked to identify the employer (or if they are not being stored on the main contractor's premises, any other person to whose order they are being held) and their destination as the works;
- there must be a written contract for the supply of the goods and materials which provides either that property passes to the purchaser unconditionally or no later than them being completed and set apart (this is known as ascertainment under the rules as to passing of property in the Sale of Goods Act 1979);

- where the goods have been ordered by a sub-contractor, the sub-contract must also be in writing and provide that as soon as property passes to the sub-contractor it immediately passes on to the main contractor;
- if the sub-contractor is manufacturing or assembling the goods, there must be a written sub-contract which provides that property will pass to the main contractor immediately on ascertainment as described above;
- the materials must be free from defects;
- the main contractor must have provided the architect with reasonable proof that he has property in the goods and materials and that the requirements listed above have been satisfied;
- the main contractor must have provided the architect with reasonable proof that the goods and materials are insured against loss or damage due to the specified perils until they are delivered to the works.

This is an extremely restrictive set of conditions, as virtually all contracts for the sale of goods and materials will contain a retention of title clause prohibiting the passing of property to the purchaser until full payment, often of all outstanding amounts rather than just those in respect of a particular consignment, has been received.

It must also be appreciated that even if the contractor and his sub-contractors are able to satisfy these conditions, payment for off-site goods and materials remains a matter for the architect's discretion, so he need not include them in an interim certificate unless he wishes to do so.

Domestic sub-contract DOM/1

The payment provisions of DOM/1 are quite different to those of NSC/C, and in many ways more important, as the arrangements for payment in most of the other standard sub-contracts are based upon them. They also mystify many sub-contractors and main contractors, because they do not reflect the practices adopted on the majority of domestic sub-contracts. Regardless of the terms of the sub-contract, many main contractors pay their domestic sub-contractors when they receive money in respect of that domestic sub-contractor's work from the employer – in other words they operate on a pay-when-paid basis. But the payment clauses in DOM/1 establish a system for paying the sub-contractor which is completely independent of certification under the main contract or receipt of money by the main contractor from the employer.

Once this basic principle has been grasped, the clauses become much easier to understand.

Clause 21 of DOM/1 provides that the first payment becomes due to the sub-contractor within one month of him starting work on site or, if so agreed, off site. Subsequent interim payments become due at monthly intervals thereafter, and payments must be made within 17 days of the due date. Thus if the sub-contractor started on site on 1 July, his first payment would become due no later than 1 August, and should be made by 18 August. Subsequent monthly payments would become due on 1 September, 1 October, etc.

This may bear no relationship at all to the payment pattern under the main contract, but there is an opportunity for synchronisation if the parties agree that the first interim payment will become due on a date within a month of the sub-contractor starting work on site which will fit in with the certification dates under the main contract.

Applications for payment

As under NSC/C, domestic sub-contractors under DOM/1 are not obliged to apply for payment, although in practice they almost invariably do so, and where the sub-contract work is of a specialist nature and therefore difficult for the average main contractor to value, this is a sound commercial approach. Where such an application is made, clause 21.4.4 of DOM/1 requires the sub-contractor to provide any details which are reasonably necessary to substantiate it.

Commercial convenience should not, however, be confused with contractual requirements, and the main contractor cannot refuse to pay the sub-contractor because he has failed to submit an application in a particular month, unless the sub-contract conditions themselves have been amended to make such an application a condition precedent to payment.

Amounts due

The amount due to the domestic sub-contractor in each interim payment is calculated in accordance with the same rules applied by the architect in determining the gross valuation of sums due to nominated sub-contractors described in relation to NSC/C. Domestic sub-contractors are entitled to be paid for goods and materials off site where the architect has exercised his discretion to certify them under the main contract, provided the domestic sub-contractor has observed the requirements of clause 30.3 of JCT 80 described earlier in this chapter.

The rates and prices used to value the domestic sub-contract works will be those agreed between the sub-contractor and the main contractor and, in theory at least, will have nothing to do with the rates and prices the main contractor is charging the employer for that work.

DOM/1 does not expressly provide who is to value the domestic sub-contract works. They will of course be valued by the quantity surveyor when he values the whole of the works under the main contract, but the figure he includes in that valuation in respect of each domestic trade will not be identified, and in any event the valuation will be at the main contractor's rates and prices, not at the sub-contractor's.

In practice, the main contractor will value the domestic sub-contractor's work and will pay him the sum he considers to be due. If the sub-contractor wishes to dispute that valuation and cannot resolve the matter by negotiation, his only remedy will be arbitration. It has recently been confirmed by the courts that an adjudicator appointed under clause 24 of DOM/1 to determine a dispute on set-off does not have the power to value the sub-contract works (*A. Cameron Ltd* v. *John Mowlem & Co. plc* (1990) 52 BLR 42).

As under NSC/C retention, amounts previously paid and cash discount at two and a half per cent should be deducted from interim payments due to domestic sub-contractors. The deductions should be made in that order to ensure that discount is deducted from the net amount due. DOM/1 is as explicit as the nominated form in providing that the discount should be deducted only where the sub-contractor has been paid on time, but given the extent to which the payment cycle under DOM/1 is ignored by all parties, proving such late payment can be extraordinarily difficult.

The additional comments on cash discount made above in relation to NSC/C are equally relevant here.

Retention

The main contractor is entitled to deduct retention of 5 per cent from payments to domestic sub-contractors unless a different percentage is inserted in the appendix to the contract. A footnote in DOM/1 states that the retention percentage should not exceed that provided for in the main contract, but the parties can ignore this advice if they so wish.

Half of the retention fund should be released on practical completion of the domestic sub-contract works. This may be considerably earlier than practical completion of the main contract works, and will result in the main contractor having to fund that payment himself, as he will not be entitled to obtain the release of any retention from the employer until the main contract works have reached practical completion.

The remaining portion of the retention will not be released to the domestic sub-contractor until the defects liability period under the main contract has expired, or the certificate of completion of making good defects has been issued under the main contract, whichever is later. There is no provision for early final payment and release of retention to domestic sub-contractors equivalent to those for nominated sub-contractors, and so early sub-contract trades can have an extremely long wait indeed before recovering the second half of their retention. It can also be difficult for them to establish the date when it is finally due, as the main contractor is not obliged to notify them when either the defects liability period expires under the main contract or when the certificate of completion of making good defects is issued.

DOM/1 does not give retention money trust status, and therefore in the event of an insolvency the domestic sub-contractor will be an ordinary, unsecured creditor in respect of his retention and will receive the same as any other unsecured creditor, which may be only a few pence in the pound.

Domestic sub-contract DOM/2

The rules in DOM/2 which provide for interim payment to domestic sub-contractors where the main contract is the JCT 81 Design and Build Form are the same as those in DOM/1.

Named sub-contract NAM/SC

The payment provisions of NAM/SC generally follow those in DOM/1, but there are two distinguishing features which arise out of clauses unique to the main contract, IFC 84.

Percentage withheld

Rather than being entitled to deduct 'retention', IFC 84 and NAM/SC permit the employer or main contractor to pay only 95 per cent of the value of work properly executed and goods and materials on or off site up to practical completion, and 97½ per cent thereafter. Although it is not described as retention, the 5 per cent withheld fulfils the same function and is subject to similar rules.

Off-site goods and materials

Payment for off-site goods and materials is at the architect's discretion under the main contract, but IFC 84 does not stipulate any criteria which have to be fulfilled before that discretion can be exercised. In practice, most architects apply similar rules to those set out in clause 30.3 of JCT 80 to ensure that their clients will obtain a good title to any off-site goods and materials for which they have paid.

Domestic sub-contract IN/SC

The provisions for interim payment in IN/SC follow those of NAM/SC.

Works Contract/2

Works contractors under the JCT Management Contract are treated in a similar way to nominated sub-contractors under JCT 80. When the architect issues interim certificates under the Management Contract, he must direct the management contractor as to the amount certified in favour of each works contractor. The management contractor must then pay the sum directed to each works contractor within 17 days of the date of issue of the certificate, less a cash discount of two and a half per cent provided payment is made on time.

The rules regarding applying for payment, the composition of payments, off-site goods and materials and cash discount explained earlier in this chapter in relation to the nominated sub-contract NSC/C are equally applicable to Works Contract/2.

Retention of 3 rather than 5 per cent will be deducted from payments due to both the management contractor and works contractors, but otherwise it is subject to the same rules as nominated sub-contractors' retention under JCT 80.

Interim payment under non-standard sub-contracts

The payment provisions are amongst the most frequently amended clauses of the standard forms of sub-contract. Sometimes the payment period itself will be extended, for example the 17 day payment period in DOM/1 may be increased to 28 days. But more radical amendments are common, and main contractor's own forms are often on a pay-when-paid basis or include no express provision for interim payment at all.

Pay-when-paid clauses

Pay-when-paid clauses are a constant source of friction between main contractors and sub-contractors. The main contractors claim they cannot afford to act as the bankers for the industry, and therefore payments to their sub-contractors must relate to the money they are receiving from the employer under the main contract. Sub-contractors argue that their contract is with the main contractor, and as they cannot control what happens at main contract level, they must be able to claim payment from the main contractor even if he has not received the money from his client.

In practice, many main contractors operate on a pay-when-paid system whatever the terms of the sub-contract may say, and provided the flow of money from the employer remains steady, this does not create any problems. The difficulty arises where money is withheld under the main contract, often for reasons entirely unconnected with the sub-contractor's performance. Will a pay-when-paid clause permit the main contractor to withhold money from the sub-contractor in those circumstances? On the face of it this is clearly the intention of such a clause, but it is not clear whether that would be its effect.

Surprisingly, the English courts have never been called upon to construe a pay-when-paid clause, but its effect has been considered in several foreign cases. In *Brightside Mechanical & Electrical Services Group Ltd* v. *Hyundai Engineering & Construction Co. Ltd* (1988) 41 BLR 110 the High Court of Singapore considered the effect of the following clause:

> 'within five days of receipt by the Contractor of the sum included in any certificate of the Architect the Contractor shall notify and pay to the Sub-Contractor the total value certified therein in respect of the Sub-Contract Works'.

The employer failed to pay sums certified under the main contract because of delay in completion, and the main contractor therefore refused to pay the sub-contractor sums certified in his favour. The sub-contractor applied for summary judgment, and the main contractor relied on the clause quoted above. As the main contractor was required to do no more than show that he had an arguable defence to defeat the sub-contractor's application for summary judgment, the court was not required to reach a firm conclusion as to the effect of the pay-when-paid clause. The judge did, however, express the provisional view that its effect was that money did not become due to the sub-contractor until it had been received by the main contractor from the employer, and therefore the sub-contractor's application for summary judgment failed.

A similar conclusion had been reached by the Hong Kong courts in the earlier case of *Schindler Lifts (Hong Kong) Ltd* v. *Shui On Construction Co. Ltd* (1984) 29 BLR 95. Although neither of these cases would bind the English courts, they would be of persuasive authority, and it is therefore quite possible that pay-when-paid clauses may mean exactly what they say.

However, most commentators believe that the main contractor can only use such a clause to delay making payment to a sub-contractor, not to refuse to pay at all. If the employer is insolvent and as a result the main contractor will never receive sums properly due to him, it is likely that the sub-contractor would be entitled to payment a reasonable time after he had satisfactorily completed his work. This is the generally held view, but it should be stressed that there are no decided cases on the point and therefore the law is not absolutely certain.

No express right to interim payment

If the sub-contract does not contain any express terms entitling the sub-contractor to receive interim payments, there may nevertheless be an implied term to that effect. For example, in *The Tergeste* [1903] P. 26 it was held that:

> 'a man who contracts to do a long and costly piece of work does not contract, unless he expressly says so that he will do all the work, standing

out of pocket until he is paid at the end. He is entitled to say "there is an understanding . . . that you are to give me from time to time, at reasonable times, payments for work done.'"

However, it might be argued that a sub-contract containing no rights to interim payment is an entire contract, that is one where complete performance by one party (i.e. satisfactory completion by the sub-contractor of all his work) is a condition precedent to any liability being imposed on the other party. If this is correct, the main contractor could refuse to pay anything at all until completion of the sub-contract works.

Although the courts will not generally find that an agreement is an entire contract unless it contains clear words to that effect, this is a risk of which sub-contractors should be aware when tendering for contracts which do not include an express right to payment on an interim basis.

Fluctuations

During the course of the sub-contract there will always be a possibility that the sub-contractor's labour and materials costs will increase due to circumstances beyond his control – imposition of new taxes, for example, or the general effects of inflation. On occasion, they may also decrease, as when employer's national insurance contributions were reduced. Some sub-contracts provide for such changes to be entirely at the sub-contractor's risk, which is referred to as a fixed price contract. Others permit them to be reflected in the sum payable to the sub-contractor, and the clauses which deal with this are known by a variety of names: contract price adjustment clauses (CPA clauses); variation of price clauses (VOP clauses) or fluctuations clauses.

With the exception of NAM/SC and IN/SC, all of the JCT-related standard forms of sub-contract provide for three possible bases on which fluctuations may be recovered:

- tax fluctuations
- conventional or full fluctuations
- formula fluctuations.

The sub-contracts for use with the Intermediate Form only provide for the recovery of tax or formula fluctuations.

The parties must agree the basis on which fluctuations are to be payable at tender stage, and the applicable clause should be identified in the contract appendix (or, where there is a standard form of tender such as NSC/T, in that tender document). If no selection is made, the sub-contract states that tax fluctuations only will be recoverable, but it is not unusual for this provision to be deleted, thus making the sub-contract totally fixed price.

Tax fluctuations

Where the sub-contractor is entitled to be paid tax fluctuations, he will be able to recover additional costs incurred due to any change in the rate or type of any contribution, levy or tax which occurs after the base date, a date defined in the tender documents or appendix which should be roughly the date on which the sub-contractor finalised his tender price.

VAT and CITB levy are expressly excluded from the definition of contributions, levies and taxes, and therefore if they increase the sub-contractor will have to bear the additional costs himself. If they should decrease, however, the sub-contractor will not be required to pass on the benefit of his lower costs.

Only changes in cost due to a change in the **rate or type** of tax will be recoverable. For example, if labour costs increase due to a wage settlement, the employer's national insurance contribution will also increase, but this will not be because of a change in the rate or type of tax, and therefore will not give rise to a right of recovery under the tax fluctuations clause.

Conventional or full fluctuations

Conventional or full fluctuations clauses permit the sub-contractor to recover a sum very close to the actual variations in his labour and materials costs. In times of low inflation they are hardly ever used, and even when prices are likely to increase substantially they are relatively rare because they are difficult and cumbersome to operate.

Where full fluctuations are recoverable, if the actual cost of employing labour increases after the base date due to an alteration in the decisions or agreements of the National Joint Council for the Building Industry or other relevant wage fixing body, the net amount of the increase will be payable to the sub-contractor.

Special arrangements apply to the recovery of fluctuations in the cost of staff employed on the site: the sub-contractor is not entitled to recover actual increases in the cost of employing such people, but he will receive an additional amount in respect of site staff when there is an increase in the cost of employing operatives.

Where fluctuations are recoverable in respect of operatives, each member of the sub-contractor's staff on site will be treated for the purposes of the fluctuations clause as though he/she were an operative employed at the highest rate, and an appropriate amount will be added to the fluctuations payment made to the sub-contractor. This additional payment may bear little relation to any actual changes in the costs of employing site staff, but will go some way towards recompensing the sub-contractor for any increased costs.

To fall within the ambit of this clause, site staff must be employed on site for at least two whole working days per week. The fluctuations payment in respect of site staff is proportionate to the amount of time each one spends on site. For example, if a contracts manager spends three days a week on site, the sub-contractor will be entitled to recover three fifths of the amount he would receive in respect of an operative paid at the highest rate in respect of that contracts manager.

Fluctuations in respect of materials costs are based on a list of market prices supplied by the sub-contractor at tender stage. The contractor is entitled to recover any increase in the market price which occurs after the base date and before the date he purchases the materials. Changes in the cost of electricity and other fuels are also recoverable, provided they are consumed on site for the purposes of the execution of the works, and are included in the list mentioned previously.

Where a full fluctuations clause applies to the sub-contract, any increased costs which the sub-contractor incurs due to changes in the rate or type of contribution, tax or levy will also be recoverable.

General rules on tax and conventional fluctuations

Both the tax and full fluctuations clauses contain certain general rules relating to the recovery of fluctuations, which are as follows:

Treatment of sub-contractors

The main contractor is obliged to include in domestic sub-contracts similar provisions for the recovery of fluctuations to those which apply to the main contract. This is frequently ignored, and is impossible for the sub-contractor to enforce, as the obligation arises under a term of the main contract to which the sub-contractor is not, of course, a party.

The mandatory forms of sub-contract − NSC/C, NAM/SC and Works Contract/2 − all include a choice of fluctuations clause, and it will be for the architect to specify the basis on which fluctuations will be recoverable in the sub-contract invitation to tender.

Notices

The sub-contractor must give a written notice to the contractor within a reasonable time of any event giving rise to a payment or recovery under the fluctuations clauses. It was confirmed in the case of *John Laing Construction Ltd* v. *County & District Properties Ltd* (1982) 23 BLR 1 that the issue of such a notice is a condition precedent to the right to recover fluctuations.

Freezing of fluctuations

Neither tax nor formula fluctuations are recoverable due to an event occurring during a period of delay for which the sub-contractor is not entitled to an extension of time. However, this 'freezing' provision should operate only where the clause entitling the sub-contractor to an extension of time is unamended, and where the contractor has responded to all of the sub-contractor's applications for extensions of time.

Daywork

Neither tax nor conventional fluctuations apply to any work which is valued on the basis of daywork.

Percentage addition

The amount recovered by the sub-contractor under the fluctuations clauses will not fully compensate him for the additional costs he will incur because they do not cover such matters as plant costs and head office overheads. There is therefore a facility in the appendix (or standard form of tender where applicable) for the parties to agree and insert a percentage addition to be added to fluctuations payments.

Where a percentage is inserted, it will usually be at a fairly low level such as 5 per cent; often it will be specified as nil or deleted altogether.

Formula fluctuations

If contractors and sub-contractors are entitled to recover anything more than tax fluctuations, they are usually calculated on the basis of a formula. The formulae used on JCT contracts were developed in the mid-1970s by a Steering Group of the Economic Development Council. They produced two types of formulae for use on building contracts: a formula for general building work based on work categories, and a number of formulae for certain specialist trades, including mechanical and electrical services, lifts and structural steelwork.

All of the formulae operate by measuring the movement of various indices, which are based on national averages of the cost of labour and materials. For this reason, the payments made under the formula may bear little relationship to the additional costs incurred by the sub-contractor, but there is generally an element of 'swings and roundabouts' in that under-recovery in one month is likely to be balanced by over-recovery in another. In any event, the simplicity of the formula compared to the administrative burden of quantifying and proving actual increases in cost ensures its continuing popularity.

Where formula fluctuations apply, a set of Formula Rules will be incorporated by reference into the sub-contract; a consolidated set of Formula Rules applying to all the standard forms of sub-contract (including the domestic forms produced by BEC, FASS and CASEC) was published by the JCT in 1987.

The Formula Rules provide that the following items are excluded from adjustment under the formula:

- work valued on the basis of daywork;
- imported goods (in respect of which the sub-contractor is entitled to recover actual fluctuations in price provided they have been listed in the sub-contract appendix or standard form of tender);
- any work valued at fair rates and prices current when the work is carried out;
- direct loss and expense.

On local authority contracts there is a facility for the employer to stipulate a non-adjustable element of up to 10 per cent, which will be deducted from formula fluctuations payments made to the contractor or sub-contractor.

The freezing provisions described above in relation to tax and conventional fluctuations apply equally to formula fluctuations.

Set-off

If the main contractor suffers loss and expense due a sub-contractor's default, he may be able to deduct that loss and expense from money which would otherwise be due to the sub-contractor. This simple concept, known variously as a set-off, cross claim, counterclaim or contra charge, is probably the source of more litigation between sub-contractors and main contractors than any other aspect of sub-contracting.

The ability to set-off claims, and thus withhold payment of sums which would otherwise be due to sub-contractors, is an extremely powerful weapon for main contractors. It gives them a significant cash flow advantage over sub-contractors, and thus puts them in a strong negotiating position. This is recognised in the standard forms of sub-contract by the inclusion of express provisions which remove the main contractor's common law rights of set-off and prescribe a procedure which must be followed if the main contractor wishes to deduct loss and expense from payments due to the sub-contractor.

It is important to appreciate that, no matter how justified the main contractor's claim for loss and expense may be, he is not entitled to set it off from sums due to the sub-contractor unless he follows the set-off procedure in the sub-contract in every detail. If he fails to do so, he must pay the sub-contractor and then pursue his claim in the ordinary way. This principle was confirmed by the Court of Appeal in *Tubeworkers Ltd* v. *Tilbury Construction Ltd* (1985).

The procedures in the standard sub-contracts are not identical, and in view of their importance they are described separately.

Nominated sub-contract NSC/C

Clause 4.27 of NSC/C provides that where the sub-contractor does not agree to the making of deductions from sums otherwise due, the following criteria must be satisfied:

(i) the main contractor must have suffered or incurred the loss (prior to Amendment 4 to NSC/C he must have **actually** incurred it – comments on the effect of this change are made later);

(ii) the set-off must have been quantified in detail and with reasonable accuracy by the contractor;

(iii) the contractor must have given the sub-contractor written notice of his intention to set-off the quantified amount and of the grounds on which the set-off is being made;

(iv) that notice must be given at least three days before the issue of the relevant certificate;

(v) if the set-off relates to delay by the sub-contractor, the architect must have issued a certificate on non-completion against that sub-contractor.

The extent to which the current wording of NSC/C allows the main contractor to set-off sums which he has not yet been forced to spend is now open to question. There were two cases on the pre-Amendment 4 requirement that the loss and expense must be 'actually incurred': *Redpath Dorman Long Ltd* v. *Tarmac Construction Ltd* (1981) CLD 07-32 and *Chatbrown Ltd* v. *Alfred McAlpine Construction (Southern) Ltd* (1986) 35 BLR 44. In both of these cases it was held that future losses, including additional preliminaries costs which would not be incurred until the original completion date for the contract had passed, could not be set-off because they had not been 'actually incurred'.

Now that the word 'actually' has been deleted, it is debatable whether this is still the position. The judge at first instance in the *Chatbrown* case laid particular stress on the word actually, and said that it gave the phrase its temporal connotation. However, a respectable argument can still be mounted that the main contractor does not 'incur' loss until he is forced to expend additional money, although there are, as yet, no decided cases on this aspect of the new wording.

Domestic sub-contract DOM/1

The set-off rules in DOM/1, are similar but not identical to those in NSC/C. The main difference is that there is no requirement for an architect's certificate of delay against the sub-contractor where the set-off relates to late completion. However, the main contractor must still satisfy the following criteria:

(i)   he must have suffered or incurred the loss;
(ii) he must have quantified the amount set-off in detail and with reasonable accuracy;
(iii) he must have given the sub-contractor written notice of his intention to set-off the quantified amount and the grounds on which the set-off is claimed to be made;
(iv) that notice must be given at least three days before the relevant payment is due.

The points made in relation to NSC/C regarding the deduction of future losses are equally relevant to DOM/1.

Although these rules will undoubtedly protect domestic sub-contractors from spurious set-offs in many circumstances, an important limitation on this protection has been highlighted by two recent Court of Appeal decisions.

In *Acsim (Southern) Ltd* v. *Danish Contracting & Development Co. Ltd* (1989) 47 BLR 55, a case involving the old 'Blue' form of domestic sub-contract, the court held that the set-off procedure was not applicable where the contractor was challenging the sum due to the sub-contractor on the basis that the sub-contractor's work had been overvalued or contained defects.

An earlier case, *B W P (Architectural) Ltd* v. *Beaver Building Systems Ltd* (1988) 42 BLR 86, had indicated that the set-off provisions contained the 'exclusive machinery' which the main contractor had to use if he wished to challenge any payment due to a sub-contractor. In *Acsim*, the Court of Appeal overruled this decision, and confirmed that the set-off rules did not apply where sums were never 'due' to the sub-contractor, for example because they related to defective work.

This principle was further extended by the Court of Appeal in the case of *A. Cameron Ltd* v. *John Mowlem & Co. plc* (1990) 52 BLR 42, in which it was held that an adjudicator's award in favour of the sub-contractor could not be enforced in the courts if the main contractor could establish that it had not been paid because no further sums were 'due' to the sub-contractor because of over-valuation or defective work.

Although the principle enshrined in these two cases is clearly correct: namely that setting off damages or loss and expense is different from disputing whether sums are due to the sub-contractor in the first place, they do make it relatively easy for the main contractor to circumvent the set-off rules by alleging that the reason for non-payment of the sub-contractor is defective work or over-valuation rather than some other default.

Domestic sub-contract DOM/2

The set-off rules in DOM/2 are precisely the same as those in DOM/1.

Named sub-contract NAM/SC

The set-off provisions of NAM/SC are precisely the same as those in DOM/1.

Domestic sub-contract IN/SC

The set-off rules in IN/SC are the same as those in DOM/1.

Works Contract/2

The set-off rules in Works Contract/2 are based on those in NSC/C, but contain some unique features due to the nature of the management contracting process. Where the management contractor wishes to set-off loss and expense which has not been agreed by the sub-contractor, under clause 4.33 of Works Contract/2 he must:

(i)   have quantified his loss in detail and with reasonable accuracy;
(ii) have given the works contractor written notice of the quantified amount together with the grounds for the set-off;
(iii) have given that notice at least three days before the relevant certificate was issued;
(iv) if the set-off relates to delay, the management contractor must have reached a decision on all outstanding applications for extension of time, and must have issued a notice of non-completion against the works contractor under clause 2.11.

Note that there is no requirement for the management contractor to have incurred the loss and expense which he wishes to deduct. This has presumably been omitted because under the relief provisions in the Management Contract it will normally be the employer who ultimately bears any loss and expense caused by the works contractor's default. The management contractor may therefore set-off any loss and/or expense and/or damage for which he 'has a claim' against the works contractor, which gives him a very wide scope to deduct money from works contractors.

Common law rights of set-off

If the sub-contract is silent on the issue of set-off, the general law will imply a right to withhold payment in certain circumstances.

For example, if the sub-contractor claims payment for work executed but the main contractor can demonstrate that he has a counterclaim for damages for late completion of that work by the sub-contractor, it is likely that the two claims are so closely connected that it would be manifestly unjust to allow the sub-contractor to enforce his right to payment without taking into account the counterclaim, and so an equitable right of set-off would exist.

In addition to this equitable right of set-off, if the main contractor is sued for payment, and he can show that the work for which the sub-contractor is claiming payment is defective, he will be able to 'set-off' his loss in diminution of the amount claimed. This is sometimes described as an abatement, which is perhaps a more accurate description than set-off because it is a separate and different right to the equitable right of set-off described earlier.

Some non-standard sub-contracts include a clause permitting claims on one sub-contract to be set-off against sums due on another. In the absence of such an express term, this would not allowed under the general law, unless the two contracts were so closely and directly related that it was almost accidental that they were not a single contract.

Where two contracts fall short of this very stringent requirement but are nevertheless closely linked, a defendant can apply to the court to have execution of any judgment on a claim under one contract stayed pending resolution of a counterclaim on another contract. The court has a discretion to order a stay in these circumstances, and cases such as *Anglian Building Products Ltd* v. *W. & C. French (Construction) Ltd* (1972) 16 BLR 1 indicate that they may be unwilling to exercise it where the plaintiff can demonstrate he is financially capable of meeting the counterclaim.

Adjudication

All of the standard forms of sub-contract covered in this book provide for the quick resolution of disputes on set-off by means of adjudication. An adjudicator can reach a decision on the basis of written evidence only, which will bind the parties to the sub-contract unless they choose to challenge it in arbitration or before the courts.

Adjudication can be an effective way for a sub-contractor to challenge what he believes to be a spurious set-off, or one where the set-off procedure has clearly been ignored. It does, however, depend upon the sub-contractor acting quickly, as he must comply with strict time limits if he wishes to invoke the procedure.

Some doubts have also been cast upon its efficacy by the case of *Cameron* v. *Mowlem*, in which it was held that an adjudicator's award could not be enforced by the courts as if it were an arbitrator's award (which, it is submitted is clearly correct) but also, more worryingly, that the main contractor need not pay any sum awarded by the adjudicator if he can demonstrate that he is refusing to do so on the grounds of a dispute as to the valuation of the sub-contract work.

However, in the later case of *Drake & Scull Engineering Ltd* v. *McLaughlin & Harvey plc* (1992) 60 BLR 102, a decision of an adjudicator that certain sums should be paid to a trustee stakeholder pending arbitration was enforced by the courts, who rejected arguments put forward by the main contractor that the adjudicator had exceeded his jurisdiction by failing to exclude undisputed set-offs and that the courts should not become involved when an arbitration was underway. The robust attitude adopted by the judge in the *Drake & Scull* case should be of considerable comfort to sub-contractors, although it will not assist them where the main contractor can demonstrate that any sums awarded to the sub-contractor are not 'due' because of earlier overvaluations or defective work.

The adjudication procedures in the different forms of sub-contract are virtually identical, and their key elements are as follows:

- the sub-contractor must give notice of adjudication to the main contractor within 14 days of receiving the main contractor's notice of set-off (this can be a major stumbling block if the main contractor has ignored the set-off procedure altogether and has failed to issue such a notice – in those circumstances it is submitted that the sub-contractor has no power to refer the dispute to adjudication);
- in that notice he must set out why he disagrees with the set-off and give particulars of any counterclaim he may have against the main contractor;
- at the same time, he must request the adjudicator to act and supply him with a copy of the above notice, the main contractor's notice of set-off, brief particulars of the sub-contract, and a copy of the architect's certificate of delay if relevant;
- the sub-contractor must also give notice of arbitration to the main contractor (although this does mean he will be forced to go to arbitration if it can be resolved by agreement or if both parties are willing to accept the adjudicator's decision);
- if the sub-contractor has made a counterclaim against the contractor, the contractor has 14 days to submit a defence to that counterclaim to the adjudicator (with a copy to the sub-contractor) – this is the contractor's only opportunity to put his case to the adjudicator and some sub-contractors avoid making a counterclaim to ensure that the main contractor will be judged solely on his original notice of set-off;
- the adjudicator must give his decision within 7 days of receiving a defence to counterclaim from the main contractor, or if none is submitted, within 7 days of the expiry of the time limit for its receipt.

The adjudicator is therefore required to make a decision very quickly and on the basis of limited written evidence. He may not call for further statements or require the parties to appear before him, although he may ask for written clarification of any ambiguities in the statements made to him by the contractor or sub-contractor. Against that background, it is not surprising that the Court of Appeal in *Cameron v. Mowlem* described an adjudicator's decision as 'ephemeral and subordinate in character' compared to an arbitrator's award, or that they made it clear that an adjudicator was not entitled to decide upon the correctness of the valuation of the sub-contractor's work.

An adjudicator has specific and limited powers under the sub-contract. He may:

- order the amount set-off should be retained by the main contractor; *or*
- order the amount set-off to be paid to the sub-contractor; *or*
- require the money to be paid to a trustee stakeholder pending arbitration or agreement between the parties; *or*
- make an order which is a combination of these.

Because of the time constraints of adjudication and the limited evidence available, orders to pay the disputed sum to a trustee stakeholder are relatively common. Although this will be important to the sub-contractor if there are serious doubts as to the main contractor's solvency, if the sub-contractor was attempting to improve his cashflow by going to adjudication, such an order may be a Pyrrhic victory.

The identity of the adjudicator and trustee stakeholder should be agreed between the parties to the sub-contract and recorded in either the standard tender document or the sub-contract appendix as appropriate. This used to mean that where no contract documentation was completed, or where phrases such as 'to be agreed' of 'from the BEC's list of adjudicators' were inserted, the adjudication provisions were almost impossible to operate. This has now been remedied by amendments to all of the JCT related sub-contracts (e.g. Amendment 3 to DOM/1 issued in 1987) which provide that if no adjudicator is named, or if he is unable or unwilling to act, the sub-contractor may unilaterally appoint an adjudicator from the BEC list. If no trustee stakeholder has been named, the adjudicator may nominate any deposit taking bank to act in that capacity.

Although these amendments have made adjudication a more accessible remedy, after *Cameron* v. *Mowlem* it is an unwieldy weapon which is relatively easy for the main contractor to evade.

Final payment

If the rules on interim payment in the JCT standard forms were followed to the letter, there should be very little money outstanding by the time the parties come to settle the final account. But, in practice, difficult issues like claims for loss and expense or payment for work which the contractor claims is a variation but the employer thinks was included in the original contract sum are often not seriously broached until final account stage, which means that (at least in the contractor's opinion) substantial sums may be due following the issue of the final certificate.

The timing of final payment is therefore important to all contractors, and perhaps most important of all to those sub-contractors who complete their work early on in the life of the project, such as steelwork or piling. However, only two forms of sub-contract, the nominated form NSC/C and Works Contract/2 provide for early final payment of sub-contractors; under all of the other standard forms the timing of the final payment under the sub-contract is linked to the final certificate under the main contract.

The timetable for final payment under each of the standard forms of sub-contract is as follows:

Nominated sub-contract NSC/C

Within six months of practical completion of the sub-contract works, the sub-contractor must send to the main contractor (or, if he is so instructed, to the architect or quantity surveyor) all of the documents necessary for the computation of the final account.

Within three months of receipt of these documents, the architect or quantity surveyor must prepare a statement showing the final sum which the sub-contractor is entitled to be paid. Sub-contractors often assume that the sum shown in that statement must be agreed with them, but this is incorrect. There is no procedure or requirement for agreement in NSC/C, and although negotiations usually take place, if no agreement can be reached the architect (or quantity surveyor) can simply issue the statement showing the sum he considers to be due, and the sub-contractor's only remedy will to be challenge it in arbitration. (Once the final certificate has been issued even that right will be lost if the arbitration notice is not issued within 28 days – comments on the conclusive nature of the final certificate follow later.)

Although the nominated sub-contractor should know the amount of his final payment within nine months of completing his work, he will only be able to claim payment of it reasonably quickly if he is able to take advantage of the provisions for early final payment in clause 35.17 of JCT 80. The operation of this clause depends upon the execution of the employer/nominated sub-contractor agreement NSC/W. Where that agreement is in place, the architect can certify the final payment due to the sub-contractor once his work is practically complete, and he must do so twelve months after such practical completion. This is subject to two provisos: the sub-contractor must have remedied all patent defects in his work and must also have provided all of the information necessary for the computation of his final account.

Notwithstanding any early final payment, the nominated sub-contractor remains liable under clause 2.12 to rectify any defects in his work until the expiry of the defects liability period under the main contract, and will of course remain liable in damages for defective work until the expiry of the relevant limitation period.

If he refuses to rectify defects during the main contract defects liability period, the architect must issue an instruction under clause 35.18 nominating a new sub-contractor to carry out that work. The employer must attempt to recover the sum paid to the new nominated sub-contractor from the original sub-contractor as damages for breach of NSC/W. If there is a shortfall between what the employer recovers and the price charged by the new sub-contractor, it will be borne by the main contractor. This means that if the original sub-contractor is insolvent the main contractor carries the risk, but he has a measure of protection against the employer accepting an unnecessarily high quote for the rectification work because his agreement to the price charged by the new sub-contractor is required, although it must not be unreasonably withheld.

If the nominated sub-contractor is unable to secure early final payment, the timing of his final payment will be related to the issue of the final certificate under the main contract. JCT 80 provides that 'so soon as is practicable', but not less than 28 days before the issue of the main contract final certificate, the architect must issue a certificate including the finally adjusted sums due under all the nominated sub-contracts. The nominated sub-contractor will then be entitled to be paid the sum included in that certificate in respect of his work in the normal way, i.e. within 17 days of its issue.

Domestic sub-contract DOM/1

Under clause 21 of DOM/1, not later than four months after practical completion of the sub-contract works, the sub-contractor must provide the main contractor with all the information he requires to calculate the final payment due to the sub-contractor. That payment will not, however, become due to the sub-contractor until seven days after the architect issues the final certificate under the main contract, and must be made within 28 days of the due date.

The main contractor must notify the sub-contractor in writing of the amount of his final payment before the due date, and must send that notice by registered post or recorded delivery. As under NSC/C, there is no requirement for the sum to be agreed with the domestic sub-contractor before the notice is issued, and if the sub-contractor disagrees with the sum notified his remedy will be arbitration.

Domestic sub-contract DOM/2

The provisions for final payment of sub-contractors employed on the basis of DOM/2 are quite different to the arrangements in DOM/1. This is because, unlike JCT 80, the JCT Design and Build Form 1981 does not require the issue of payment certificates by an architect, but provides for payment on the basis of applications submitted by the contractor. Although the timing of the final payment to the domestic sub-contractor under DOM/2 is linked to the final payment due from the employer under the main contract, the language used in DOM/2 is quite different from that in DOM/1, and the sub-contract must also cater for some complicated fall-back provisions in JCT 81 which stipulate the procedure to be followed where the contractor fails to apply for his final payment at the proper time.

Under clause 21.8.1 of DOM/2, the sub-contractor must provide the main contractor with all the documents necessary for the calculation of the final payment within two months of the practical completion of the sub-contract works.

If the sub-contractor fails to do this, the contractor may give him notice that, unless he provides the relevant documents within the next two months, the contractor himself may ascertain the final sub-contract sum. This is a parallel provision to that in the main contract which allows the employer to value the final account if the main contractor does not submit his application on time.

If the final sub-contract sum is ascertained by the contractor in accordance with this clause, it will become final and conclusive of the amount due to the sub-contractor unless it is disputed in writing by the sub-contractor within one month of the date of its issue.

The final payment becomes due to the sub-contractor either 14 days after the main contractor's final account and final statement have become conclusive as between the main contractor and employer, or 14 days after the employer's final account and final statement have become conclusive.

As under DOM/1, before the due date the main contractor must notify the sub-contractor in writing of the amount of his final payment by registered post or recorded delivery.

The final payment must be made within 21 days of becoming due.

Named sub-contract NAM/SC

The final payment provisions of NAM/SC are the same as those in DOM/1.

Domestic sub-contract IN/SC

The final payment provisions of IN/SC are the same as DOM/1.

Works Contract/2

The rules as to final payment to works contractors under Works Contract/2 are the same as those in NSC/C, save that there is no provision for mandatory early final payment to works contractors. If, however, a works contractor requests early final payment and that request is approved by the employer or the architect acting on his behalf, final payment to that works contractor may be included in an interim certificate issued under the Management Contract. Clause 8.4 of the Management Contract states that such early final payment is subject to the works contractor having satisfactorily indemnified the management contractor against any latent defects.

Effect of final certificate/final payment

All of the JCT main contracts covered in this book contain similar provisions to the effect that the final certificate is final and conclusive as to certain matters, and the standard forms of sub-contract contain similar provisions in relation either to the final certificate or to the final payment.

Taking JCT 80 as an example, the final certificate is conclusive evidence that:

(i)   where the quality of materials or the standard of workmanship are to be to the reasonable satisfaction of the architect, that they are to his reasonable satisfaction;
(ii)   effect has been given to all the terms of the contract relating to the calculation of the final payment, save for accidental or mathematical errors;
(iii) all extensions of time have been awarded;
(iv) all loss and expense has been reimbursed.

The final certificate will not, however, be final and conclusive as to the matters raised in any litigation or arbitration commenced within 28 days of its issue. This requires the actual issue of a writ or notice of arbitration; measures which fall short of this such as mentioning the existence of a dispute in correspondence, threatening legal proceedings or instructing solicitors are not sufficient.

The time at which the final certificate or payment has this effect varies slightly as between the various sub-contract forms:

• under NSC/C, it is 21 days from the date the final certificate is issued under the main contract;
• under DOM/1, DOM/2, NAM/SC and IN/SC, it is either ten days after the final payment is made, or ten days after the receipt by the sub-contractor of the notice of the amount due, whichever occurs first;
• under Works Contract/2, it is 28 days after the issue of the final certificate under the Management Contract.

One aspect of the conclusive effect of the final certificate and final payment which has often caused confusion is the extent to which it can relieve the contractor of responsibility for defective work. Until recently, it was generally thought that its effect in this regard was extremely limited, and that it applied only to materials or workmanship which were expressly required by the contract documents to be to the architect's reasonable satisfaction.

But in the case of *Colbart Ltd* v. *H. Kumar* (1992) 59 BLR 89, it was held that the relevant clause in IFC 84 applied not only to anything stipulated in the contract documents to be to the architect's reasonable satisfaction, but also to anything which was **inherently** a matter for his opinion. This naturally raises the question of what is inherently a matter for the architect's opinion, and unfortunately the judgment in the *Colbart* case gives little assistance in that regard.

This decision has been criticised by many commentators, and was not followed in the recent case of *Darlington Borough Council* v. *Wiltshier Norther Ltd* (1993) (unreported) (although the latter case concerned the construction of the equivalent provision in JCT 63 and was distinguished on those grounds). If *Colbart* v. *Kumar* is correct, the final certificate or final payment could relieve both main contractors and sub-contractors of liability for many types of defect. Although this may sound an attractive proposition for contractors, it is likely to cause many architects to be extremely reluctant to issue final certificates, and it is therefore hoped that the JCT will clarify its contracts on this point in the near future.

Remedies for late and non-payment

Most of the remedies available to a sub-contractor in the event of late or non-payment are considered elsewhere in this book: arbitration, litigation and ADR are examined in Chapter 12, 'Methods of Dispute Resolution'; determination is covered in Chapter 11 on 'Determining a Sub-Contractor's Employment'; deduction of cash discount and adjudication are covered elsewhere in this chapter.

Two particular remedies for non-payment are particularly linked to the payment provisions themselves and are therefore covered here: direct payment of nominated sub-contractors and suspension of work.

Direct payment

Only nominated sub-contractors employed under JCT 80 are entitled to direct payment from the employer in the event of non-payment by the main contractor. It will be mandatory only where the nominated sub-contractor has entered into the employer/nominated sub-contractor agreement NSC/W, otherwise it is purely discretionary.

A detailed procedure governing the amount and timing of direct payments to nominated sub-contractors is set out in clauses 35.13 of JCT 80. Before the issue of each interim certificate under the main contract, the main contractor must provide the architect with reasonable proof that he has paid the sums certified in favour of the nominated sub-contractor in the last certificate (or that he is not obliged to do so, for example because he has made a set-off in accordance with the procedure in NSC/C).

If he cannot provide that proof, the architect issues a certificate to that effect, with a copy to the nominated sub-contractor. The employer then deducts the unpaid sum from the next payment due to the main contractor, and pays it direct to the nominated sub-contractor. The direct payments are therefore always one month behind the payments the sub-contractor should be receiving, and are therefore not the panacea many sub-contractors imagine.

The employer is not bound to pay any more than is due to the main contractor, and so is not at risk of being out of pocket. If direct payments are due to several nominated sub-contractors but there is not enough money to pay all of them in full, the employer may either pay them pro rata or divide up the available funds on any other basis he considers to be fair and reasonable.

A very important limitation on the protection offered by the direct payment provisions is that they cease to operate on the winding up or bankruptcy of the main contractor – the very time when the sub-contractor is most in need of them. The Tribunal was forced to introduce this restriction to take account of the case of *British Eagle International Airlines Ltd* v. *Compagnie Nationale Air France* [1975] 2 All ER 390. In that case, the House of Lords held that any contractual arrangement which had the effect of reducing the sum available to all creditors by paying it directly to an unsecured creditor was void as being contrary to the Companies Acts and public policy.

However, direct payments may still be made where the main contractor is involved in an insolvency procedure which falls short of liquidation or bankruptcy, such as administration or administrative receivership.

Direct payment of retention money to nominated sub-contractors should also be possible, even where the main contractor is being wound up or made bankrupt, because it is trust money and therefore does not form part of the main contractor's ordinary assets.

Suspension of work

Stopping work is the single most powerful weapon available to a sub-contractor who is not being paid in accordance with his sub-contract. The enormous disruption which can be caused by a sub-contractor who withdraws his labour from the site at a vital point in the programme is generally enough to concentrate the mind of the most recalcitrant main contractor, and will also have the added bonus of bringing the sub-contractor's complaint to the attention of the employer and the professional team.

It is a remedy which must, however, be used with great care. A sub-contractor who suspends work when he is not contractually entitled to do so is repudiating his sub-contract, which will entitle the main contractor to employ another sub-contractor to complete the work, and to recover all of the additional costs of doing so (including the costs of delay and disruption to other trades) from the repudiating sub-contractor.

It is also of limited assistance to those sub-contractors who work on site for only a short period, as non-payment to them often does not become apparent (or demonstrable) until after they have completed their work.

It should be noted that main contractors do not have the option of suspending work if they are not paid by the employer; they may determine their employment altogether, but they are not contractually permitted to withdraw from site on a temporary basis.

The procedures for suspending work in the different forms of sub-contract vary, and each of them is therefore described below.

Nominated sub-contract NSC/C

A nominated sub-contractor may suspend work 35 days after the issue of an interim certificate in respect of which the main contractor has not paid or otherwise discharged his liability to the sub-contractor (e.g. by a valid set-off).

The 35 day period has been stipulated to give the employer an opportunity to operate the direct payment provisions if appropriate, and thus avoid the disruption of the sub-contractor's withdrawal.

Before suspending work, the sub-contractor must give 14 days written notice to both the main contractor and the employer; this can be done as soon as the 17 day period within which the main contractor should have paid has elapsed.

The sub-contractor must not suspend work 'unreasonably or vexatiously', so a minor non-payment such as the wrongful deduction of cash discount in a particular month would probably not entitle him to suspend.

Domestic sub-contract DOM/1

Under clause 21.6 of DOM/1 the sub-contractor may withdraw from site in the event of a breach of the payment provisions which continues for seven days after the sub-contractor has issued a written notice of that breach to the main contractor.

Although the remedy can be implemented very quickly under DOM/1, in practice it can be difficult to prove that there has been a genuine breach of the payment provisions. This is because, in practice, it is the main contractor who values the domestic sub-contract work, and so the sub-contractor must be very sure that he is genuinely entitled to more than he has been paid before he withdraws his labour. If there is a total non-payment this may be a simple matter, but if the sub-contractor feels that variations are being consistently undervalued, suspension may be quite a high risk strategy.

Domestic sub-contract DOM/2

The provisions of DOM/2 on suspension of work are the same as those in DOM/1.

Named sub-contract NAM/SC

The rules in NAM/SC governing suspension of work are the same as those in DOM/1, save that the sub-contractor must copy his notice of intention to suspend to the architect.

Domestic sub-contract IN/SC

IN/SC contains exactly the same provisions regarding suspension as DOM/1.

Works Contract/2

Works contractors may suspend work if they are not paid by the management contractor in accordance with Works Contract/2, provided they give seven days written notice to the management contractor. As they are paid on the basis of architect's certificates they do not usually have the problems suffered by domestic sub-contractors of having to prove how much was due to them.

8 Progress and Extensions of Time

Some of the fiercest disputes on building contracts arise out of delays in completion. More often than not, all of the various parties will be involved: employer, professional team, main contractor and sub-contractors, all vigorously blaming each other for the delay whilst proclaiming their own innocence. There are usually only two incontrovertible facts: that the building is late and that it will cost someone a great deal of money. The key questions are, of course, why, who and how much? The purpose of this chapter is to answer those questions in relation to sub-contracts.

Establishing the sub-contract period

Before questions of delay and extensions of time can be properly considered, it is necessary to establish when the sub-contractor will be obliged to complete his work, assuming that everything goes according to plan. Unlike the JCT standard forms of main contract, which provide for a date of possession and a date for completion to be agreed and inserted in the contract appendix, all of the JCT related sub-contracts provide for the agreement of a sub-contract period.

Where a standard form of tender such as NSC/T or NAM/T is used, the sub-contractor is required to specify in his tender:

- the period of notice he requires to start work on site;
- the period of time he requires to execute any off-site work;
- the period of time he requires to carry out his work on site.

That information is submitted to the main contractor when the sub-contractor is named or nominated, and it is one of the issues which must be agreed between main contractor and sub-contractor before the sub-contract is executed. It is often a major bone of contention, and may require an instruction from the architect to, for example, extend the main contract completion date before it can be resolved.

Exactly the same information is included in the appendix to the domestic sub-contracts. When they submit their tenders, domestic sub-contractors should specify the periods of time they require to execute their work both off site and on site, and the period of notice they require to commence work on site. These should then be agreed with the main contractor before the sub-contract is finalised.

All too frequently, sub-contractors fail to follow this basic rule, and enter into a sub-contract where the period for execution of the work on site is 'to be agreed' or 'to suit main contractor's programme'. Leaving matters such as the sub-contract period to be agreed at a later date may be tempting when there are so many other pressing matters to deal with, but it can severely prejudice the sub-contractor.

Unless an agreement is eventually reached (and the sub-contractor will be unable to insist upon this, as agreements to agree are generally held to be unenforceable) the sub-contractor will be required to complete his works within a reasonable time. As a person obliged to complete within 'a reasonable time' has been held by the courts to have fulfilled that obligation:

'notwithstanding protracted delay, as long as such delay is attributable to causes beyond his control, and he has acted neither negligently nor unreasonably' (*Hick v. Raymond & Reid* [1893] AC 22)

this might seem like an attractive proposition for the sub-contractor. The problem is that, without going to court on the issue, it will be extremely difficult for the sub-contractor to establish what a reasonable time would be in any given circumstances. The main contractor is likely to act as though the sub-contractor is bound by his programme, and to set off any losses he suffers if the sub-contractor does not comply with the programmed dates. However contractually incorrect this may eventually prove to be, it can seriously disrupt the sub-contractor's cashflow.

Agreeing to carry out the work 'to suit main contractor's programme', without a basic entitlement to a minimum period of weeks for the execution of the work on site, can also be dangerous for the sub-contractor.

In *Martin Grant & Co. Ltd v. Sir Lindsay Parkinson & Co. Ltd* (1984) 29 BLR 31 sub-contractors employed on the basis of a non-standard form of contract agreed to 'proceed with any portion . . . of the work at such time . . . as the contractor should require having regard to the requirements of the contractor in reference to the progress or conditions of the main works.' The court held that such words were sufficient to negative any implied term that the main contractor would do nothing to prevent the sub-contractor from completing his works, and therefore the sub-contractors were not entitled to recover loss and expense from the main contractor on the grounds of disruption. In effect, the sub-contractors had agreed to execute their work in a way which would suit the main contractor, regardless of any inconvenience it would cause to them.

Although it is debatable whether agreeing to carry out the sub-contract works 'to suit the main contractor's programme' in the context of a standard form of sub-contract, with its elaborate provisions for extensions of time and loss and expense, would have quite such a drastic effect, there is no doubt that it would weaken any claim the sub-contractor wished to make for disruption costs.

It is important to note that the sub-contract period is usually calculated by reference to a number of weeks rather than to fixed dates. This means that, although the sub-contractor will be entitled to the notice stipulated in the sub-contract before he can be required to commence work on site, when he enters into the sub-contract the dates on which he will be required to start and finish his work may not be known.

All of the JCT related sub-contracts require the sub-contractor to:

- carry out and complete the sub-contract works in accordance with the relevant details in the tender document or appendix; *and*
- reasonably in accordance with the progress of the works (i.e. the main contract works); *but*
- subject to receipt of the notice to commence work on site as detailed in the tender document/appendix and to the operation of the extension of time clauses.

This may mean that the sub-contractor will be required to visit the site on a number of separate occasions, and provided that is a consequence of carrying out his works reasonably in accordance with the progress of the rest of the project, the mere fact that he has to do so will not entitle him to claim disruption costs. The only exception to this rule is where the sub-contractor has made it clear that he has based his price on a maximum number of visits to the site and this has been reflected in the sub-contract appendix or tender document.

For those trades which involve a significant amount of off-site fabrication, such as structural steelwork and lifts, the right to an agreed period of notice before they are contractually required to start work on site will be crucial to their ability to carry out their work within the sub-contract period. Strictly speaking, that notice period should be sufficient to allow for the execution of any off-site work, therefore the period may need to be considerably longer than the usual one or two weeks.

It is in the interests of all parties to ensure that the agreed periods are achievable, as both over-optimistic predictions by sub-contractors and unrealistic demands by main contractors generally result in a disillusioned client and a series of claims which could have been avoided.

Time is not of the essence

No standard form of sub-contract provides for time to be of the essence. The effect of such a term would be that if the sub-contractor was in breach of his obligations relating to completion, the main contractor would be entitled to treat the sub-contract as being at an end, and would be discharged from continuing to perform his part of the bargain. This would be the case no matter how minor the delay, as by making time of the essence the parties would be agreeing that any delay went to the root of the contract and thus gave the injured party (the main contractor) the right to treat it as repudiated.

If there is no express term that time is of the essence, such a term will not be implied in a contract which contains provisions for extensions of time and the payment of liquidated damages in the event of delay.

If both an express term making time of the essence and the machinery of extensions of time and liquidated damages are present, it is debatable whether the courts would allow the main contractor to treat the sub-contract as repudiated in the event of any delay beyond the sub-contract period as extended by any extensions of time. In *Peak Construction (Liverpool) Ltd* v. *McKinney Foundations Ltd* (1971) 1 BLR 111 it was suggested that this would be the result, but the remark was obiter dicta (i.e. not crucial to the decision in that case and therefore not binding in future cases) and therefore the point remains undecided.

Status of programmes

There is often considerable confusion over the status of the various programmes which are produced by the parties to the building process. This is because although they undoubtedly have a vital part to play in the management of any project, they may be of little contractual effect.

Under the standard JCT main contracts, it is clear that the main contractor is bound only to complete by the completion date (that is the original date for completion stated in the contract as extended by any extensions of time) and is not obliged to complete any particular operation by a given date. This was confirmed by the courts in the case of *M. J. Gleeson (Contractors) Ltd* v. *Hillingdon Borough Council* (1970) 215 EG 165, in which it was held that because the JCT main contracts provide that nothing in the bills of quantities or similar documents will override or modify the interpretation of anything in the contract conditions, a detailed programme in the bills showing dates when parts of the building should be completed did not bind the contractor.

The position in relation to sub-contracts is not quite so straightforward, as it clearly vital that the work of all the various trades on site is properly coordinated, and the most efficient way to achieve this is by a detailed programme produced by the main contractor. However, the starting point must always be that neither the main contractor nor the sub-contractor is bound to perform in accordance with a programme unless compliance with it is an express contractual obligation.

All of the standard forms of sub-contract for use with the JCT forms require the sub-contractor to carry out and complete his work in accordance with the details set out either in the standard tender document or in the sub-contract appendix. If those details include a detailed programme, the sub-contractor will be contractually bound to perform in accordance with that programme. If they do not, he will free to carry out his work in any order he chooses, provided he completes within the sub-contract period and works reasonably in accordance with the progress of the main contract works.

Often, a programme will be included in the sub-contract documents or referred to in the bill of quantities or specification, rather than in its proper place, namely the sub-contract appendix or standard tender document. The status of such programmes is highly questionable, and it is therefore unclear whether the sub-contractor will be in breach of contract if he fails to complete a particular operation by the date shown on the programme. This is because all of the JCT related sub-contracts contain a clause which provides that, in the event of a conflict between the sub-contract conditions and the numbered documents, the sub-contract conditions will prevail. The sub-contract conditions make it quite clear that the proper place for any detailed programme requirements with which the sub-contractor must comply is either the standard form of tender if the sub-contractor is nominated under JCT 80 or named under IFC 84, or the sub-contract appendix if a standard domestic form is being used. It could therefore be argued with some force that the inclusion of a programme anywhere else creates a 'conflict', with the result that only those details in the tender document or appendix are binding.

A detailed programme which is contractually binding may, of course, be of as much value to the sub-contractor as the main contractor, as it will generally place obligations on both parties, not just the sub-contractor. If, therefore, it provides that a certain area will be made available to the sub-contractor on a particular date and it is not, due to some fault of the main contractor or another sub-contractor, the sub-contractor will have a right to claim damages or loss and expense. Even if the programme is not legally binding, if the sub-contractor is prevented from working in accordance with it, that will be good evidence that 'the regular progress of the Sub-Contract Works has been materially affected', thus opening the door to a claim under the sub-contract for loss and expense for disruption.

Whatever the contractual status of a programme, it is almost inevitable that it will require some revision during the progress of the project to take into account variations and delays which have already occurred. The question of whether those revisions will bind the sub-contractor will depend upon the status of the original programme. If it was included or referred to in the sub-contract appendix or the standard tender document, revisions will not be contractually binding unless the sub-contractor agrees to them. It should, however, be borne in mind that he will have a continuing obligation to work in accordance with the progress of the works as a whole; therefore he cannot simply seek to adhere to an programme which does not take into account the fact that there have been major variations or substantial delays.

If the programme itself makes it clear that the main contractor has the power (or even the duty) to vary it unilaterally to take into account the actual progress of the project, then such variations will be as binding as the original programme.

Notice requirements

All of the sub-contracts for use with the JCT standard forms impose detailed requirements for notices of delay to be issued by the sub-contractor to the main contractor. Although compliance with those requirements has been held by the courts in *London Borough of Merton* v. *Stanley Hugh Leach Ltd* (1985) 32 BLR 51 not to be a condition precedent to the sub-contractor's right to be awarded an extension of time, failure to issue the relevant notices will, in practice, weaken the sub-contractor's position in any negotiations regarding delay.

No distinction is drawn in this section between the different standard forms of sub-contract, as they contain virtually identical provisions in this regard.

Time of issue

Whenever it becomes reasonably apparent that the commencement, progress or completion of the sub-contract works is, or is likely to be delayed, the sub-contractor must issue a notice of delay 'forthwith'. It will usually become apparent that there may be some delay as soon as certain events occur – for example, when a major variation is instructed or when severe weather conditions are encountered. 'Forthwith' is defined by the *Concise Oxford Dictionary* as 'immediately, without delay', and therefore a notice issued several weeks after such an event had occurred would not satisfy the requirements of the sub-contract.

Notice should be given when the actual progress of the sub-contract work has been delayed, not the hoped for or expected progress. If the sub-contractor had planned to complete his work within six weeks of a ten week sub-contract period, it is only when an event occurs which is likely to delay him beyond the ten week period that a notice of delay should be issued. This point was confirmed by the courts in the case of *Glenlion Construction Ltd* v. *The Guinness Trust* (1987) 39 BLR 89.

Content of notice

The notice must be in writing, but there is no requirement that any particular form of words should be used. The information which should be contained in the notice is, however, stipulated in the sub-contract, and there are several matters which should be covered in a properly drafted notice of delay.

Firstly, the sub-contractor should set out the material circumstances including, insofar as the sub-contractor is aware of them, the causes of the delay. Often the sub-contractor will be able to identify only the immediate cause, for example that he is being delayed by late completion of work by others. He will often not know why that other work has been delayed – it could be late information from the architect, lack of coordination by the main contractor or default by another sub-contractor. In such circumstances the sub-contractor would be well advised not to speculate or allocate blame, he should simply record the facts of which he is aware, and leave it at that.

Secondly, the sub-contractor should identify any matter which in his opinion amounts to either a 'Relevant Event' (which are listed later in this chapter) or to an act, omission or default of the main contractor. Sometimes this will be straightforward, for example if there has been an event such as a fire on site or a strike. However, in the case of delays which may have been caused by others, the sub-contractor should be sure of his facts before stating that, for example, a delay is due to late information from the architect or to the main contractor's default.

The sub-contractor should give particulars of the expected effect of each of the delaying events identified in his notice. This requirement is often ignored, although the information will be of considerable importance to the person responsible for awarding the extension of time. It will involve explaining the precise effect the event will have on the progress of the sub-contractor's work – whether it involves a complete cessation of work, for example, or whether it simply involves disruption because of the need to divert labour from one operation to another.

Finally, the sub-contractor should estimate the length of the delay. Sub-contractors (and, for that matter, main contractors) are often reluctant to include an estimate of delay in their initial notice, preferring to state simply that they will use their best endeavours to prevent any delay from occurring, and giving an estimate only at a later stage, often after the original completion date or sub-contract period has expired. This will rarely be a good tactic. An architect or main contractor who has been told that there will probably be no delay at all may decide not to take any remedial measures (such as re-programming or the omission of certain work) on the strength of that assurance, and will be much less amenable to requests for extensions of time at a later date. If anything, the estimate of delay in the original notice should be slightly pessimistic, as it can easily be revised in a subsequent notice when more information becomes available.

Further notices

Having issued the initial notice when the delaying event occurs, many sub-contractors take no further action until they wish to make a formal claim for an extension of time towards the end of their sub-contract. This is incorrect: the sub-contractor is under an obligation to issue further written notices updating the particulars and estimates he originally gave as further information becomes available.

Events entitling the sub-contractor to an extension of time

Sub-contractors will be entitled to an extension of time if they are delayed by:

- a Relevant Event as defined in the sub-contract; *or*
- an act, omission or default of the main contractor, his servants, agents or other sub-contractors.

The Relevant Events in the various forms of sub-contract are similar but not identical, and so each sub-contract is commented upon in turn.

Nominated sub-contract NSC/C

Force majeure

This is a term from French law which literally means 'superior strength'. It is sometimes incorrectly described as 'Act of God', but in fact it is much wider than that as it includes man-made events such as strikes and embargoes. The most frequently quoted definition is to be found in the case of *Lebeaupin* v. *R. Crispin & Co.* [1920] 2 KB 714, in which McCardie J. approved the following definition taken from French law:

> 'This term is used with reference to all circumstances independent of the will of man, and which it is not in his will to control . . . thus, war, inundations and epidemics are cases of *force majeure*; it has even been decided that a strike of workmen constitutes a case of *force majeure*.'

The term has also been considered by the English courts in the case of *Matsoukis* v. *Priestman* [1915] 1 KB 681, when Bailhache J. found that the general coal strike of 1912 was *force majeure*, as was a breakdown of machinery. The latter point may be particularly useful to sub-contractors who are involved in off-site (or on-site) fabrication involving complicated machinery.

Some commentators have suggested that *force majeure* has a more restricted meaning within the standard forms of building contract because certain matters such as strikes, fire and weather are dealt with separately; it could equally be argued that the relevant events are not mutually exclusive in the sense that a flood, for instance, could constitute both *force majeure*, exceptionally adverse weather conditions and a specified peril.

Exceptionally adverse weather conditions

The weather in question must have an adverse effect on the operations of the sub-contractor who is claiming the extension. It must also be exceptional for the time of year and the area in which the work is being carried out. A measure of bad weather is to be expected, particularly during the winter months, and sub-contractors will be expected to take this into account when negotiating their sub-contract period.

If necessary, the sub-contractor should be prepared to demonstrate that the weather is exceptional by reference to local weather records: airports can be a good source of these if there is no local meteorological office.

In *Walter Lawrence* v. *Commercial Union Properties* (1984) 4 Con LR 37 it was held that the effect of the weather was to be assessed at the time the work was actually carried out, not when it was programmed to be carried out, but this has subsequently been questioned in an *obiter dicta* comment in *Balfour Beatty Building Ltd* v. *Chestermount Properties Ltd* (1993) 62 BLR 1.

Loss or damage caused by the specified perils

The sub-contractor has the benefit of the insurance policy taken out by either the main contractor or employer in respect of certain defined risks known as 'the Specified Perils'. The occurrence of loss or damage caused by any of those specified perils will also entitle the sub-contractor to an extension of time.

The specified perils are defined as follows:

- fire
- lightning
- explosion
- storm
- tempest
- flood
- bursting or overflowing of water tanks, apparatus or pipes
- earthquake
- aircraft and other aerial devices dropped therefrom
- riot and civil commotion

- but excluding the excepted risks, which are nuclear risks and sonic booms.

It is sometimes argued that a sub-contractor is not entitled to an extension of time if the specified peril has been caused by his own negligence. This is most likely to be an issue in the case of fire or water damage. The relevant event itself is not subject to a qualification that the specified peril must not have been caused by the sub-contractor's negligence, so it would appear from the face of the contract that the sub-contractor will be entitled to an extension of time.

However, the right of the architect or main contractor to grant an extension of time to the sub-contractor is subject to a proviso that the sub-contractor shall constantly use his best endeavours to prevent delay to the works, and this is hardly consistent with an entitlement to an extension where it is the sub-contractor himself who has caused the delaying event. Against this, it can be argued that the whole basis of the provisions dealing with specified perils is that there should be no investigation into whether they were caused by the negligence of any party, but that the consequent damage should simply be remedied as quickly as possible.

On balance, it is suggested that the latter view is to be preferred, but sub-contractors should be ready for a long battle if they are claiming an extension for a delay caused by rectifying damage due to a specified peril which arose from their own negligence.

Strikes, etc.

This relevant event is not confined to official strikes – it includes civil commotion, local combination of workman and lock-out. However, if an industrial dispute falls short of an actual strike, in that it does not involve a stoppage of work, but simply disrupts progress due to working to rule or other obstructive activities, it is unlikely that the sub-contractor will be entitled to an extension.

The strike must affect a trade employed on the works themselves, or employed in the preparation, manufacture or transportation of any materials or goods required for the works, and would therefore encompass a haulage strike or a strike at a manufacturer's factory.

A strike affecting the sub-contractor's own labour would be covered, although if the strike had been caused by a deliberate act of the sub-contractor, such as failing to follow an applicable collective agreement, it may be argued that the sub-contractor has failed to exercise his 'best endeavours' to prevent the delay.

Architect's instructions

The sub-contractor will be entitled to an extension of time if he is delayed by complying himself or by the main contractor complying with architect's instructions in relation to the following matters:

- a discrepancy or divergence in or between the main contract documents or the nominated sub-contract numbered documents;
- variations;
- expenditure of provisional sums (unless the instruction relates to the expenditure of a provisional sum for defined work, in which case both *SMM 7* and the contract require the contractor to have taken the programming implications of executing the relevant work into account when compiling his tender);
- postponement of any work to be executed under the contract;
- antiquities;
- nominated sub-contractors under clause 35 (although if a nominated sub-contractor's employment is determined and the architect takes

only a reasonable time to nominate a replacement, it is thought that there will be no entitlement to an extension of time);
- nominated suppliers under clause 36;
- opening up and testing of any work, materials or goods, unless the relevant work, materials or goods are not in accordance with the contract.

Late instructions, etc.

This is probably the most contentious of all the relevant events as lack of information, particularly relating to design, is a problem on many projects, and may imply a default by one of the parties involved in awarding the extension, namely the architect.

The nominated sub-contractor will be entitled to an extension of time if either he or the main contractor is delayed because they have not received from the architect in due time any necessary instructions (including instructions relating to the expenditure of provisional sums), drawings, details or levels. This right is subject to either the contractor or sub-contractor having specifically applied in writing for the relevant information 'on a date which having regard to the Completion Date or the period or periods for the completion of the Sub-Contract Works was neither unreasonably distant from nor unreasonably close to the date on which it was necessary for the Contractor or Sub-Contractor to receive the same'.

Many contractors and sub-contractors submit an 'information required schedule' to the architect at the beginning of the contract, which is often in the form of their programme, annotated with the dates upon which they will require specific information. The status of such a schedule was considered by the courts in *London Borough of Merton* v. *Leach*, in which it was held that the schedule did satisfy the requirement for written application for information, and that although the schedule was submitted at the beginning of the contract period, it was not submitted on a date 'unreasonably distant' from the date when the contractor actually required the information.

However, if such a schedule is to be used, it should be regularly updated to take account of any variations and/or delays which occur during the course of the contract, and it would also be prudent for the contractor and sub-contractor to supplement it by the issue of specific notices requesting information at the appropriate time.

An important restriction on the sub-contractor's right to an extension of time on this ground is that the date on which the architect is required to supply the information should be assessed by reference to the completion date, or the period for completion of the sub-contract works. This means that if either the main contractor or sub-contractor has programmed the work so that it will be completed before the relevant date or the expiry of the relevant period, they will be unable to force the architect to provide them with any outstanding information to enable them to do so.

This principle was confirmed in the case of *Glenlion Construction Ltd* v. *The Guinness Trust*, in which a main contractor failed in his claim for disruption costs which he had incurred because the architect had failed to give him information at a time which enabled him to complete his work within his own programme, which was considerably shorter than the contract period.

Nominated sub-contractors and nominated suppliers

If a nominated sub-contractor is delayed by any other nominated sub-contractor or by a nominated supplier, he will be entitled to an extension of time. The reason for the delay by the nominated sub-contractor or supplier is irrelevant: regardless of whether it is caused by the nominated sub-contractor's own negligence or by some cause beyond his control, both the main contractor and all other sub-contractors will be entitled to an extension of time as a result, provided the main contractor took all practicable steps to avoid or reduce the delay. If the delay was caused by the nominated sub-contractor's own default, the employer will be left to seek a remedy from the nominated sub-contractor under the employer/sub-contractor warranty NSC/W.

This relevant event has a somewhat restricted meaning following the decision in *J. Jarvis & Sons* v. *Westminster Corporation* (1970) 7 BLR 64, in which it was held that it did not extend to a delay caused by a nominated sub-contractor returning to site to remedy defects in his work, but only to delays he caused during the time he was originally on site to carry out the nominated sub-contract work.

Execution of work or supply of materials by employer

The employer is permitted by clause 29 of JCT 80 to engage his own labour to work on site during the contract period, although if the contract documents do not provide sufficient information about such direct works to allow the contractor to fulfil his own contractual obligations, the employer must obtain the contractor's consent. If the execution of such other work causes a delay, both the contractor and all of his sub-contractors will be entitled to an extension of time.

The wording of the relevant event is sufficiently broad to encompass any statutory undertakers employed under a direct contract with the employer. If, however, statutory undertakers are simply undertaking the work because they are legally obliged to do so, any delay they may cause will fall within the relevant event relating to statutory undertakers. This is an important distinction, as disruption due to employer's work will entitle the contractor and his sub-contractors to loss and expense, but delay due to statutory undertakers carrying out their statutory obligations will not.

If the employer has chosen to provide materials and goods for the execution of the work, either at a price or on a free issue basis, the sub-contractor will be entitled to an extension if any delay is caused as a result.

Exercise of statutory power by UK government

The sub-contractor will be entitled to an extension for any delay caused by the execution of a statutory power by the UK government after the base date, provided it affects the execution of the works by:

- restricting the availability or use of labour; *or*
- prevents or delays the obtaining of materials, goods or fuel which is essential to the proper carrying out of the works.

This might occur if the government imposed trade sanctions against a country supplying oil or some other important commodity.

Inability to obtain labour or materials

These relevant events were notorious in the 1963 edition of the JCT standard form as clauses 23(j)(i) and (ii), because they were marked with an asterisk which indicated that they could be deleted if the parties so agreed. In practice, they were hardly ever allowed to remain in the contract, and although the asterisk was not included in the 1980 version of the standard form, they are still frequently deleted from both the main contract and the related sub-contracts.

Even if they apply to the sub-contract, these grounds for extension of time are very limited indeed. The sub-contractor's inability to obtain labour or materials must not only be beyond his control, but must be for reasons he could not reasonably have foreseen at the base date. This means that if, for example, a supplier was named (not nominated) in the sub-contract documents, and the sub-contractor was well aware when he tendered for the work that the supplier was habitually delivering goods three months late due to a glut of work, he would not be entitled to an extension of time if that supplier failed to deliver on time.

Similarly, the labour or materials must be **essential** to the proper carrying out of the works. If an alternative is available, even if it is more expensive, it should be used – at no cost to the employer unless it constitutes a variation.

Local authorities and statutory undertakers

This relevant event applies only where the work is being undertaken by the local authority in pursuance of its statutory obligations. The privatisation of bodies such as British Gas does not affect the position as they continue to fulfil statutory obligations notwithstanding their change of ownership.

If the statutory undertaker is being employed by the employer, any delays he may cause should also entitle the contractor and his sub-contractors to loss and expense (see earlier section on the execution of work or supply of materials by the employer).

The statutory obligation being fulfilled must relate to the works: if a statutory undertaker impedes access to the site by working on a water main or electricity cable, the contractor and his sub-contractors will be entitled to an extension under this heading only if the water main or cable relates to the works. If not, this might qualify as *force majeure*, but that is by no means certain.

Lack of access

This does not encompass any lack of access to the site, but applies only where the employer has failed to give access to the site or any part of it which is over land in his possession or control.

Suspension by the sub-contractor

If the nominated sub-contractor validly exercises his right to suspend further execution of the sub-contract works because he has not been paid by the main contractor in accordance with the sub-contract, any consequent delay will entitle him to an extension of time.

Deferment of possession

If the optional provision allowing the employer to defer giving possession of the site to the main contractor for up to six weeks applies to the main contract, and the employer exercises that option, the nominated sub-contractor will be entitled to an extension of time.

Approximate quantities not being a reasonably accurate forecast

This applies only if bills of quantities are one of the nominated sub-contract numbered documents, and although those bills are firm, they have been prepared in accordance with *SMM 7* and therefore contain some approximate quantities. The sub-contractor will be entitled to an extension of time if he is delayed because the amount of work actually carried out means that the approximate quantity was not a reasonably accurate forecast.

Domestic sub-contract DOM/1

The relevant events in the domestic sub-contract DOM/1 are exactly the same as those in the nominated sub-contract NSC/C.

Named sub-contract NAM/SC

The relevant events in the named sub-contract NAM/SC are the same as those in sub-contract NSC/C, save that there are no provisions relating to delays by nominated sub-contractors or suppliers for the simple reason that the Intermediate Form, for use with which NAM/SC is drafted, does not provide for nomination.

In the relevant event relating to delay caused by architect's instructions there is, however, a reference to instructions issued under clause 3.3 of the main form, which relates to named sub-contractors.

If a named sub-contractor is delayed by lack of progress by another named sub-contractor, there is no relevant event applicable to that delay, but it will be treated as an act or default of the main contractor, and therefore the 'innocent' named sub-contractor will be entitled to an extension of time on that basis. (The main contractor himself will be unable to obtain an extension of time, and will therefore be liable to pay liquidated damages to the employer for the delay, which he will seek to recover from the defaulting named sub-contractor.)

However, if the named sub-contractor's default is so serious that his employment is determined, the employer will bear part of the risk of that default. If the named sub-contractor's employment is determined due to his default or insolvency, the architect must issue instructions naming another person to complete the relevant work, or for the main contractor to complete it himself, or to omit it. Whatever he decides, the main contractor and any sub-contractors affected by those instructions will be entitled to an extension of time.

If the named sub-contractor determines his own employment due to the main contractor's default, that will not fall within the relevant event, but will entitle any innocent named sub-contractors who may be affected by the delay to an extension on the grounds of the main contractor's default.

There is no relevant event in NAM/SC relating to the exercise of a statutory power by the UK government after the base date, but if this occurs, the sub-contractor would probably be entitled to an extension of time on the grounds of *force majeure*.

The relevant events regarding inability to obtain labour and materials are optional in the Intermediate Form, and will apply to the named sub-contract only where they are stated to be applicable to the main form in the tender document.

Domestic sub-contract IN/SC

The relevant events in IN/SC are exactly the same as those in the named sub-contract NAM/SC.

Domestic sub-contract DOM/2

The relevant events in sub-contract DOM/2 are similar to those in NSC/C although as the relevant main contract, the JCT With Contractor's Design Form 1981, does not cater for the appointment of nominated or named sub-contractors (save for very brief naming provisions in the supplementary provisions) there is no reference to delays by or instructions relating to nominated or named sub-contractors.

In addition, the terms used in the relevant events are slightly different to cater for the use in JCT 81 of employer's requirements and contractor's proposals rather than other contract documents. References to late instructions etc. are to late instructions from the employer rather than the architect, as JCT 81 does not provide for the appointment of an architect,only for the appointment of an employer's agent.

JCT 81 includes two relevant events which are unique to that form and its sub-contract DOM/2, namely:

(i)   delay in receiving permission or approval from statutory bodies which the contractor has taken all practicable steps to avoid or reduce; *and*

(ii) delay consequent upon a change in the statutory requirements after the base date which necessitates an amendment to the contractor's proposals or an amendment to the contractor's proposals which is necessary to take account of a decision of a planning authority.

There is no relevant event regarding the use of approximate quantities in an otherwise firm bill, as JCT 81 does not provide in any detail for the use of bills of quantities.

Works Contract/2

The scheme of the provisions governing extensions of time in the JCT Management Contract is very different to the more conventional JCT forms.

The management contractor is entitled to an extension of time if completion of the project is delayed by what is described as a 'Project Extension Item', which is either:

any cause which impedes the management contractor in carrying out his obligations including:

* a default by the employer
* late instructions from the professional team
* deferment of possession
   (it should be noted that this is a non-exclusive list);

   or

- relevant events as defined in the works contracts, save for culpable delay by other works contractors.

Although the management contractor will not be entitled to a project extension if completion is delayed by the default of any works contractor, under the relief provisions the employer may claim or deduct liquidated damages from payments due to the management contractor only to the extent that the management contractor has been able to recover them from the defaulting works contractor.

In spite of these differences, the relevant events themselves in Works Contract/2 are in very similar terms to those in NSC/C. The key differences are:

- a strike must affect a trade employed on or in the preparation, manu-facture or transportation of goods for use on the works which the works contractor himself is carrying out; it will not be sufficient if a strike affects only another area of the project;
- there is no reference to delays caused by instructions relating to nomi-nated sub-contractors as there is no such thing under the management contract (although there is provision for nominated suppliers);
- there is an additional relevant event providing for a works contractor to be entitled to an extension of time if he is delayed by another works contractor;
- the relevant events relating to inability to obtain labour or materials refer only to the works contractor's inability, not to any inability by the management contractor, in contrast NSC/C refers to the inability of either the nominated sub-contractor or the main contractor;
- there is no relevant event dealing with suspension of work by the works contractor due to non-payment, but this is a difference of form rather than substance, as the works contract makes it clear that any suspension resulting from a non-payment will entitle the works contractor to an extension on the grounds of the management contractor's default.

Procedure for awarding extensions of time

The different sub-contracts for use with the JCT standard forms vary in the way in which they allocate responsibility for awarding extensions of time to sub-contractors, and therefore each sub-contract is considered separately.

Nominated sub-contract NSC/C

The main contractor awards extensions of time to nominated sub-contractors under clause 2.3 of NSC/C, but he may not do so without the consent of the architect. The clause makes it clear that an extension will be granted only if the architect is of the opinion that the sub-contractor has been delayed either by a relevant event or by the main contractor's act, omission or default, and that it is the architect's estimate of what is a fair and reasonable extension of time which must be awarded. The main contractor's function is therefore simply an administrative one: the power of decision lies with the architect.

This arrangement has been much criticised on the grounds that the architect will often be unaware of the facts relating to the nominated sub-contractor's progress, although this is dubious given the detailed notices, particulars and estimates which the sub-contractor is required to provide when a delay occurs.

It is also said that it places the architect in an invidious position because he has to determine whether the main contractor's default has delayed the sub-contractor, and therefore become involved in 'domestic' disputes between the main contractor and his sub-contractor. However, as the alternative would be the main contractor himself making that decision from a less than impartial position, the arrangement is generally popular with sub-contractors.

When granting an extension of time the contractor must, in agreement with the architect, indicate which relevant events have been taken into account, and the extent to which regard has been had to the omission of any work.

If 'reasonably practicable', the revision to the sub-contract period should be made within 12 weeks of the main contractor having received reasonably sufficient notices, particulars and estimates from the sub-contractor, or before the end of the sub-contract period if that occurs first. The contractor is obliged to notify the sub-contractor of negative decisions of the architect, as well as positive ones, within the relevant period.

In *Temloc Ltd* v. *Errill Properties* (1987) 39 BLR 30, a main contractor sought to argue that the failure of the architect to comply with a requirement in JCT 80 to decide upon extensions of time within 12 weeks phrased in precisely the same terms as that in NSC/C meant that the liquidated damages clause was inapplicable. The Court of Appeal held, quite rightly, that the compliance by the architect with the time limits was not a condition precedent to the award of an extension of time. However, Croom-Johnson LJ went on to say that the requirement to reach a decision within twelve weeks was 'directory only', and therefore did not impose a duty on the architect. Most commentators found this view surprising, as the clause in both the main contract and its related sub-contracts uses the term 'shall' rather than 'may', which generally indicates an intention to impose a mandatory requirement. It remains to be seen whether Croom-Johnson LJ's view will be supported in subsequent cases.

Within 12 weeks of practical completion of the nominated sub-contract works, the architect must review the extensions of time previously granted to the nominated sub-contractor, and:

- extend the period for completion previously fixed; *or*
- fix a shorter period than that previously fixed (but **only** on the grounds of the omission of work, obligations or restrictions); *or*
- confirm the period for completion previously fixed.

Clause 25.3.6 of JCT 80 prevents the architect from fixing a completion date for the main contract works before the original date for completion inserted in the contract appendix, but there is no similar restriction on the architect's powers under the sub-contract, and therefore if there have been major omissions from the sub-contract works, the period fixed under the review provisions may be shorter than the original sub-contract period.

However, omissions are the only ground upon which previously granted extensions can be reduced, and therefore any award of an extension of time before the sub-contract works have been completed tends to be conservative.

If the sub-contractor is aggrieved by the architect's refusal to consent to the granting of an extension of time, or by the period of any extension granted, he may use the main contractor's name to arbitrate against the employer for a breach of the main contract by his agent, the architect, provided he gives the main contractor such indemnity and security as he reasonably requires.

This device, described as 'name-borrowing', is necessary because neither the employer nor the architect is a party to the nominated sub-contract and therefore because of the rules of privity, they cannot be sued directly by the sub-contractor for breach of contract.

Domestic sub-contract DOM/1

Under DOM/1, it is the main contractor who is given the power to award extensions of time to the domestic sub-contractor, without any participation whatsoever by the architect. There is therefore no need for the 'name borrowing' arrangements which exist in NSC/C.

Other than this crucial distinction, the procedure is as already described in relation to NSC/C, save that the time period within which decisions must be made and the review carried out is 16 rather than 12 weeks.

Named sub-contract NAM/SC

The main contractor also decides upon extensions of time due to named sub-contractors under clause 12 of NAM/SC. This is consistent with the principle that, once appointed, named sub-contractors are treated as domestic.

The procedure for awarding extensions of time is less detailed than in the JCT 80 related sub-contracts, and there is no fixed period within which the main contractor must give his decision. He must, however, act 'so soon as he is able to estimate the length of the delay'.

Although the main contractor has the power to review extensions previously granted at any time, he is not bound to do so. If he does choose to conduct a review, he may not reduce any extensions previously granted, regardless of any omissions from the works.

Domestic sub-contract IN/SC

The procedure for granting extensions of time in IN/SC is precisely the same as in NAM/SC.

Domestic sub-contract DOM/2

The procedure for granting extensions of time to domestic sub-contractors employed under DOM/2 is exactly the same as those applicable to DOM/1.

Works Contract/2

The management contractor decides upon the entitlement of works contractors to extensions of time under clause 2 of Works Contract/2. But before he either grants an extension or refuses to do so, the management contractor must notify the architect of his decision. The architect's consent is not required, and his views do not affect the works contractor's entitlement. However, he may 'dissent' from the management contractor's decision if he thinks it is incorrect, which is likely to have an effect on the management contractor's entitlement to a project extension under the Management Contract. For this reason, it will be a brave management contractor who awards an extension to a works contractor from which the architect has dissented, and it seems fair to assume that, in practice, the architect will have considerable influence over extensions of time awarded to works contractors.

Concurrent causes of delay

The task of the person responsible for awarding extensions of time to the sub-contractor is far from simple. All of the standard forms of sub-contract require that a 'fair and reasonable' extension of time is granted, which will usually involve the exercise of judgment and discretion rather than a simple calculation.

One of the most common complicating factors is that relevant events rarely occur in isolation, and therefore the architect or main contractor will be forced to confront the difficult problem of concurrent causes of delay.

Sometimes, the concurrent causes will all entitle the sub-contractor to an extension of time – for example, bad weather and late instructions. For the purpose of simply avoiding liquidated damages, it will not matter to the sub-contractor which relevant event is identified as the event giving rise to the extension, but it will be crucial in relation to his claim for loss and expense due to disruption.

On other occasions, the choice between concurrent causes may directly affect the sub-contractor's right to an extension, for example if he is delayed by both architect's instructions and his own default.

There are a number of different approaches to this issue, and at present the law is unclear as to which is correct. It is, however, suggested that the best method is to look for what has been described as 'the effective, dominant cause' of the delay. The courts have given little guidance to those faced with the task of applying that test to a given set of circumstances, beyond stating that it will be a question of fact in every case, and that it should be established by applying common sense standards, not simply by treating the most recent event as the dominant cause. (Those requiring further details are referred to *Keating on Building Contracts* (fifth edition, 1991) at pages 192–197, which contains an excellent and very full discussion of this problem.)

A similar difficulty arises where a relevant event occurs during a period of culpable delay by the sub-contractor, and therefore extends it. This has recently been considered by the courts in *Balfour Beatty Building Ltd* v. *Chestermount Properties Ltd.*

In that case, the architect issued instructions to a main contractor during a period of culpable delay which involved the carrying out of a considerable amount of extra work. The contractor argued that JCT 80 did not permit the architect to grant an extension of time in those circumstances, and therefore the instructions amounted to an act of prevention by the employer which resulted in time becoming at large. In the alternative, he argued that the architect should grant a 'gross' rather than a 'net' extension of time, thus adding the period required to carry out the extra work to the period of time the contractor had already taken when the instruction was issued (which would, of course, have the effect of wiping out the period of culpable delay).

The judge rejected both of these arguments. He held that the architect was entitled by the wording of clause 23.3.3 of JCT 80 to award an extension in respect of a relevant event which occurred after the most recently fixed date for completion of the works, and therefore time did not become at large. Further, in deciding upon the length of the extension of time, the architect should add the number of days fairly and reasonably necessary to carry out the additional work to the most recently fixed date for completion, thus awarding the extension on a net rather than a gross basis. He even went so far as to say that if the additional work could reasonably be carried out simultaneously with the existing work, the contractor might be entitled to no extension of time at all.

It is therefore fairly clear that where a relevant event which constitutes an act or default of the employer or main contractor arises during a period of culpable delay by the sub-contractor, the sub-contractor will be entitled to an extension of time for any additional delay it causes although it will not, as Balfour Beatty sought to argue, magically wipe out the culpable delay already caused by the sub-contractor.

Where the relevant event is 'neutral' – for example a strike or bad weather, it is sometimes argued that if the sub-contractor had not been in delay in the first place his progress would not have been adversely affected, and so the dominant cause of the delay is not the weather, etc., but his own default, therefore he should not be entitled to an extension. This construction was rejected by the courts in *Walter Lawrence* v. *Commercial Union Properties*, but was supported in an obiter comment (which is not therefore binding on other courts) by the judge in the *Balfour Beatty* case. The question therefore remains unresolved.

Best endeavours

All of the JCT related forms of sub-contract make the sub-contractor's entitlement to an extension of time subject to the proviso that he has used his 'best endeavours' to prevent delay in the progress of the sub-contract works 'howsoever caused'. It is debatable whether this means the sub-contractor is bound to use his best endeavours to prevent a relevant event from occurring, or whether it simply imposes a duty to mitigate any delay caused by a relevant event. The normal, natural meaning of the words suggests that both are covered, although this is contrary to the usual interpretation of the relevant event involving the specified perils which, it is thought, includes specified perils caused by the sub-contractor's negligence.

There are no decided cases on the issue, and so sub-contractors should be prepared for an uphill struggle when claiming extensions of time for any relevant event caused by their negligence.

Exactly what action the sub-contractor must take in order to fulfil the best endeavours requirement will depend on the circumstances of the case. It would encompass such measures as carrying out his work in a slightly different order, but it is generally accepted that the sub-contractor is not obliged to expend significant sums of money by working overtime.

Time at large

The phrase 'time at large' is used to describe the situation where a contractor or sub-contractor is obliged to complete his works within a reasonable time. If he fails to do so, the other party to the contract will be entitled to recover unliquidated damages for any delay, which means that he will be required both to quantify and prove his loss.

Time will become at large in the following circumstances:

(i)   where no period for completion of the works has been agreed by the parties to the contract; *or*
(ii)   where one party is under a contractual obligation to complete his work within a particular period, but completion is delayed by an act of pre-vention by the other party to the contract and any extensions of time clauses in the contract do not expressly cater for that act.

In the event of time becoming at large under the main contract, any provision for liquidated damages will simply cease to operate, and will be replaced by a right to recover unliquidated damages if the contractor does not complete within a reasonable time. This is one of the reasons why standard forms of contract contain such detailed provisions for extensions of time, and why such clauses tend to be construed strictly against the employer, because they are included essentially for his benefit to prevent time from becoming at large.

The effect of time becoming at large under a sub-contract is slightly different. This is because, strictly speaking, liquidated damages do not directly apply to the sub-contract, but are part of the unliquidated damages the sub-contractor will be required to pay in the event that he delays the main contractor in circumstances which make the main contractor liable to pay liquidated damages under the main contract.

When time becomes at large under the sub-contract, the sub-contractor's obligation to complete his work within the sub-contract period will, as under the main contract, be replaced by an obligation to complete within a reasonable time. However, the unliquidated damages the main contractor can recover if the sub-contractor fails to do so may include an element of the liquidated damages applicable to the main contract if time is not also at large under that contract.

Practical completion

The date on which practical completion of the sub-contract works occurs is of crucial importance in the JCT-related sub-contracts, because it marks the date upon which:

* liability to pay damages for delay ceases;
* the defects liability period begins;
* the first half of the sub-contractor's retention fund is released;
* the sub-contractor's protection under the main contract all risks policy ends;
* arbitration can be commenced on any issue.

Definition of practical completion

In spite of its significance, the term 'practical completion' is not defined in any of the standard forms of contract, and perhaps like Lord Denning's apocryphal elephant, it is easier to recognise than define. It is a subject upon which a number of judges have speculated, but it is extremely difficult to synthesise their various views into a single definition.

Generally the courts have taken a fairly narrow view of what constitutes practical completion which does not accord with the practice generally adopted in the industry, at least in a climate where employers actually want to occupy their buildings.

For example, in *J. Jarvis & Sons* v. *Westminster Corporation* Viscount Dilhorne said:

'One would normally say that a task was practically completed when it was almost, but not entirely finished, but practical completion suggests that that is not the intended meaning and what is meant is the completion of all the construction work that has to be done.'

The theory that a certificate of practical completion should not be issued if there are any patent defects at all in the works is also supported by the wording of JCT 80, which limits the contractor's express obligation to make good defects to those which 'appear within' the defects liability period, and therefore does not apply to any defects which may already have manifested themselves before the certificate was issued.

In *H.W. Nevill (Sunblest) Ltd* v. *William Press & Sons Ltd* (1982) 20 BLR 78 the Court took a slightly broader view of practical completion, when the judge said:

'I think the word practically . . . gave the architect a discretion to certify that [the contractor] had fulfilled its obligation under clause 21(1) (of JCT 63), where very minor *de minimis* work had not been carried out, but that if there were any patent defects in what [the contractor] had done the Architect could not have given a certificate of practical completion.'

Most of the leading authorities on practical completion were recently reviewed by His Honour Judge Newey in the case of *Emson Eastern Ltd* v. *EME Developments Ltd* (1991) 55 BLR 114. His comments on practical completion are *obiter dicta* because they are not crucial to the decision in that case, as it turned upon the meaning of the word 'completion' rather than the phrase 'practical completion'. They are nevertheless useful, and should also be comforting to contractors, as Judge Newey adopted a very pragmatic approach.

He pointed out that building construction was not like the manufacture of goods in a factory. The size of the project, site conditions, the use of many materials and the employment of various types of operatives made it virtually impossible for the contractor to achieve the same degree of perfection as a manufacturer. He also said that it must be rare in a new building for every screw and brush of paint to be correct, thus undermining by implication the principle that there must be no patent defects at all when practical completion is certified.

Perhaps the only real saving grace for contractors is that the question of whether a building is practically complete can be referred to arbitration. Unlike a judge, an arbitrator under the JCT standard forms can actually substitute his discretion for that of the architect, and therefore, if necessary, the contractor can go to arbitration to ensure that the words are sensibly construed.

Procedure for determining practical completion

The standard forms of sub-contract contain different arrangements for determining the date of practical completion, so each of the forms is dealt with in turn.

Nominated sub-contract NSC/C

Under clause 35.16 of JCT 80, the architect is obliged to issue a separate certificate of practical completion in respect of each nominated sub-contract when, in his opinion, the nominated sub-contract works are practically complete.

Domestic sub-contract DOM/1

No certificate is issued to mark the date of practical completion of domestic sub-contract works, and therefore different rules have evolved to identify the date upon which it occurs.

Under clause 14 of DOM/1, the sub-contractor must notify the contractor of the date on which he thinks that the domestic sub-contract works are practically complete. If the contractor does not dissent from that date within 14 days of receiving the sub-contractor's notice, practical completion is deemed to have occurred on the date notified by the sub-contractor.

If, however, the main contractor does so dissent, the date can either be established by agreement, or if the parties so wish, it can be referred to arbitration. If neither of these occurs, the sub-contract works will be deemed to be practically complete at the same time as the main contract works. This can create serious problems for early trades such as structural steelwork and piling, whose work may be finished months or even years before the building is practically complete. This is one of the reasons why such sub-contractors prefer to be nominated.

Named sub-contract NAM/SC and domestic sub-contract IN/SC

The rules for ascertaining practical completion under the sub-contracts for use with the Intermediate Form are exactly the same as those under DOM/1.

Domestic sub-contract DOM/2

The rules relating to practical completion in DOM/2 are exactly the same as those in DOM/1.

Works Contract/2

Under Works Contract/2, the management contractor is required to issue separate certificates of practical completion for each works contract. However, such a certificate cannot be issued until the relevant works are practically complete in the opinion of the architect, so the provisions are not dissimilar to those in the nominated sub-contract.

Liquidated damages

The sum of money which the main contractor will be required to pay to the employer for each day or week of a delay for which the contractor is not entitled to an extension of time will usually be pre-agreed between the parties and inserted in the contract appendix – in other words, liquidated damages will be payable in the event of delay.

The amount of liquidated damages inserted in the appendix will be the exact amount recoverable by the employer if the contractor is in culpable delay, even if his actual loss is very much greater or much less than the pre-agreed sum. This rule can produce extraordinary results: the employer may be suffering no loss at all if the building is a speculative development and the bottom has dropped out of the property market and yet will be entitled to his liquidated damages, but conversely if rents have doubled since he calculated the rate he inserted in the contract, he will be unable to recover his additional loss.

A liquidated damages provision will be unenforceable in certain circumstances, notably:

- if time becomes at large (as described earlier in this chapter); or
- if the relevant sum is a penalty.

The amount inserted in the contract will be treated as a penalty rather than as liquidated damages if, at the date the contract was entered into, it was not a genuine pre-estimate of the loss which the employer would suffer if a delay occurred. In the leading case of *Dunlop Pneumatic Tyre Company* v. *New*

Garage and Motor Co. Ltd [1915] AC 79 the House of Lords held that an agreed sum would be a penalty if it was 'extravagant and unconscionable' compared to the greatest possible loss that could flow from the breach, assessed at the date the parties agreed the sum, not at the date the loss occurred.

None of the standard forms of sub-contract provide for the sub-contractor to be liable to pay liquidated damages to the main contractor in the event of a delay. However, they will be part of the unliquidated damages which the contractor is entitled to recover from the sub-contractor if, due to some omission or default of the sub-contractor, the main contractor is forced to pay liquidated damages under the main contract.

This is because under the second rule in *Hadley* v. *Baxendale* damages are recoverable if they arise due to special circumstances known to the parties at the time they enter into the contract. Sub-contractors will be aware when they enter into a sub-contract that liquidated damages will usually be payable under the main contract in the event of delay, and the sub-contractor will generally also be aware of the rate of liquidated damages applicable to the main contract. This should always be the case where a standard form of tender such as NSC/T has been used, because such forms include a copy of the main contract appendix in which the rate will appear. Even if a standard form of tender has not been used, the main contractor would (and should) have informed the sub-contractor of the rate when he invites him to tender.

Where the sub-contractor is nominated under JCT 80, if he delays the main contractor the main contractor will be entitled to an extension of time under clause 25.4.7, and will not therefore be obliged to pay liquidated damages to the employer. This does not mean that the sub-contractor is 'off the hook', as he will be liable to the employer for a breach of the warranty NSC/W, in which he undertakes not to delay the contractor so that he is entitled to an extension of time.

There is some uncertainty as to how the employer's loss should be quantified in these circumstances. It is sometimes argued that what the employer has lost due to the sub-contractor's breach is the right to claim liquidated damages under the main contract, and therefore the amount of those damages is the amount he is entitled to claim. An alternative, and probably better, view is that the employer must prove and quantify his actual loss due to the delay, and therefore the amount of the liquidated damages is irrelevant. In practice, most employers simply claim the amount of liquidated damages, but if there is a substantial difference between the two figures, it may be worth an argument on the point.

Sub-contractors sometimes attempt to argue that they should not be subject to the full amount of liquidated damages applicable to the main contract, and therefore seek to limit the extent to which the main contractor can recover this element of his loss from them. This may be justified if the value of the sub-contract is very low compared to the value of the whole project, particularly if a high rate of liquidated damages has been agreed.

However, if a small sub-contractor delays such a project, the effect will be just as devastating as a delay caused by a major contractor.

For certain trades, especially in good trading conditions, the only way to secure sensible, competitive tenders from sub-contractors may be to restrict their liability to pay to the main contractor liquidated damages which he will be liable to pay in the event of delay. Few main contractors will be willing to retain this risk, and therefore such limitations tend to be accepted only where there is a corresponding agreement between the main contractor and employer which similarly limits the employer's rights in the event of a delay caused by that sub-contractor.

Acceleration

If a contract has got into delay, or if the client discovers he urgently needs his building at an earlier date than he originally anticipated, he may decide that the building works should be accelerated, either by reducing extensions of time to which the contractor and his sub-contractors would otherwise be entitled, or by adjusting the original date for completion.

None of the more traditional standard forms of contract provide for such acceleration, and so it can only be achieved by a separate agreement properly supported by consideration. Sub-contractors will obviously have a key role to play in any acceleration, and should therefore either be a party to the agreement with the client, or should enter into their own agreement with the main contractor.

The JCT Management Contract and its associated works contract are the only standard JCT forms which currently provide for acceleration. The purpose of the clauses dealing with acceleration in those contracts is to allow the management contractor and his works contractors to control both the extent to which acceleration can be achieved and the price to be paid for it.

The detailed procedure in clause 3.4 of Works Contract/2 is as follows:

(i)   The employer causes the architect to issue a preliminary instruction under clause 3.6.3 of the Management Contract requiring acceleration or an alteration in the sequence or timing of the work.

(ii) The management contractor issues that preliminary instruction to all works contractors who will be affected by it and asks whether they have any reasonable objection to complying. Any such objection must be made within 7 days of receipt of the instruction, or within such longer period as may be reasonable.

(iii) If any works contractor does object, that objection is passed to the architect by the management contractor, together with the management contractor's comments (which must also be copied to the works contractor).

(iv) If the architect decides that the works contractor's objection is reasonable, the architect either withdraws his preliminary instruction, or varies it so as to remove the objection; in the latter case the works contractor must confirm in writing that his objection is withdrawn.

(v) The onus then shifts to the works contractor, who must submit his acceleration proposals to the management contractor for onward transmission to the architect 'as soon as reasonably practicable'. The proposals should comprise either the lump sum which the works contractor requires or, if it is not reasonably practicable to state a lump sum, that the cost of compliance with the instruction should be ascertained under the loss and expense provisions. The works contractor must also state the reduction in the period for completion of his works which can be achieved, or the extent to which extensions of time which would otherwise be granted could be cancelled or reduced.

(vi) The architect may then issue an instruction under the Management Contract confirming the works contractor's proposals **in exactly the same terms** as they were submitted. He may not instruct measures which go beyond those proposed by the works contractor, or treat an offer to accelerate by one week in return for £1000 as an offer to accelerate by any given period at a rate of £1000 per week.

(vii) That instruction is issued by the management contractor to the works contractor, who must comply with it. He will be entitled to the sum put forward in his proposals and will be bound by the new completion date identified in those proposals.

(viii) The extensions of time clauses in the works contract will continue to apply, and therefore if the works contractor is delayed by a relevant event occurring after the instruction to accelerate has been issued, the works contractor will be entitled to an extension of time.

Both under the JCT Management Contract and other JCT forms where negotiations lead to a separate, ad hoc agreement, acceleration can be fraught with problems. Often, the contractor and his sub-contractors or works contractors will take the agreed acceleration measures – working overtime, carrying out operations simultaneously, paying premiums to ensure goods and materials are available – and yet will fail to achieve the accelerated completion date. Few clients will be willing to pay acceleration costs in such circumstances, although they may properly be due if the failure was due to a relevant event occurring after the acceleration had commenced.

In the context of sub-contracts, the key points to be borne in mind when negotiating an acceleration agreement are as follows:

• main contractors should consult with sub-contractors before they consent to an acceleration agreement, and if key trades are affected they should be directly involved in the negotiations with the client;

- trades involving a substantial amount of off-site fabrication such as lifts and structural steelwork may be physically unable to accelerate because of manufacturing constraints, regardless of the amount of money available to pay for the acceleration;
- the acceleration agreement should be absolutely clear as to whether the extensions of time provisions will continue to operate during the period of the acceleration if the contractor or his sub-contractors are delayed by relevant events which occur after the acceleration agreement has been concluded.

9 Sub-Contractor's Claims for Loss and Expense

In the context of building contracts, the term 'claim' is generally used to describe an application for payment for anything other than the measured value of the work originally required by the contract. The most common sources of claims are variations, delay and disruption. The provisions for valuing variations are described in Chapter 6; this chapter is concerned with the claims which can be made by sub-contractors when they incur additional costs due to either delay or disruption of the regular progress of their work.

Legal basis of claims

If a claim is to have a realistic chance of success, it must be based on sound legal principles. The mere fact that it has cost the sub-contractor more than he anticipated to carry out his sub-contract work will not be sufficient: he must be able to show either that the sub-contract entitles him to extra payment or that the main contractor is in breach of an express or implied term of the sub-contract which has directly resulted in the sub-contractor incurring additional costs.

Claims therefore fall into two categories:

(i) those made under a specific provision of the sub-contract; *and*
(ii) claims made on the grounds of a breach of the sub-contract for which there is no express provision in the contract itself, often described as 'common law claims'.

Each of these categories is considered in detail below.

Claims provided for by the sub-contract

All of the JCT related sub-contracts contain detailed provisions entitling the sub-contractor to recover loss and expense in the event that the regular progress of his work is materially affected by either

- an act, omission or default of the main contractor or his other sub-contractors; *or*
- one of the 'Relevant Matters' listed in the sub-contract.

It is important to appreciate that an entitlement to recover loss and expense under the sub-contract will not necessarily be linked to delay: what the sub-contractor must prove is that the regular progress of his work has been materially affected, colloquially referred to as disruption. An extension of time may be granted in circumstances where no loss and expense is recoverable (for example if the delay was caused due to bad weather or strikes), and loss and expense may be payable for disruption even though the sub-contractor has completed his work within the original sub-contract period.

The 'Relevant Matters' which will entitle the sub-contractor to recover loss and expense are as follows:

(i) Late instructions, etc.

If the disruption is caused by either the main contractor or the sub-contractor not having received in due time the necessary instructions, drawings, details or levels from the architect, loss and expense will be recoverable. However, this right is subject to the same proviso which applies to the relevant event entitling the sub-contractor to an extension of time, so either the contractor or the sub-contractor through the contractor must have specifically applied in writing for the relevant information at the appropriate time.

The comments regarding 'information required schedules' in Chapter 8 are therefore equally applicable here.

(ii) Opening up or testing

Disruption caused by opening up or testing will only entitle the sub-contractor to recover loss and expense in very limited circumstances, namely where:

- it has been carried out pursuant to an instruction issued under the main contract (rather than provided for in the contract documents); *and*
- that instruction was not issued as a result of the discovery of defective work or materials (save under Works Contract/2, but this will be brought into line during 1994 when new provisions on opening up and testing will be introduced); *and*
- the goods, materials or workmanship tested were found to be in accordance with the contract.

(iii) Discrepancies in or divergences between documents

If the main contractor or a sub-contractor finds a discrepancy in or conflict between the documents prepared on behalf of the employer under the main contract, such as the contract drawings and the contract bills, the architect must issue an instruction to resolve the discrepancy or conflict. If that instruction causes disruption, the sub-contractor may recover loss and expense.

(iv) Employer's work or supply of materials

The employer may have chosen to reserve certain aspects of the building work to be carried out by someone other than the main contractor; this is particularly common where the fitting out of the building is let separately from the construction of the shell and core. The presence of such other contractors on the site can cause considerable logistical difficulties for both the main contractor and his sub-contractors, but loss and expense due to any disruption they may cause will be recoverable under this relevant matter.

If the employer is supplying goods or materials for the works and fails to do so at the proper time, loss and expense will also be recoverable.

(v) Architect's instructions to postpone

Under the main contract the architect (or employer in the case of JCT 81) has wide powers to instruct the postponement of any work: if he does so both the main contractor and his sub-contractors will be entitled to recover the loss and expense they suffer as a result.

(vi) Lack of access

Disruption caused by lack of access to the site or any part of it will give rise to an entitlement to loss and expense, but only if it is due to the employer's failure to provide access over land in his possession and control. To give rise to a valid claim such access must also be either indicated in the contract documents (in which case the contractor must have given any notice to the architect required by the contract documents) or agreed with the architect.

(vii) Variations

Architect's instructions requiring variations or as to the expenditure of provisional sums for undefined work, and main contractor's directions passing on such instructions, will entitle the sub-contractor to recover loss and expense.

(viii) Approximate quantities

Assuming the bills of quantities have been prepared in accordance with *SMM 7*, if any approximate quantities are included in those bills which are not a reasonably accurate forecast of the quantity of work actually required, the sub-contractor may claim loss and expense for any delay or disruption which results.

In addition to these 'Relevant Matters' there are three other instances in which the sub-contractor will be entitled to claim loss and expense under the express terms of the sub-contract:

(i) Default of main contractor

This includes not only actions of the main contractor himself which may result in disruption to the sub-contractor, but also those of other sub-contractors or anyone else for whom the main contractor is responsible.

(ii) Deferment of possession

The main contract contains an optional clause allowing the employer to defer taking possession of the site for a period of up to six weeks: if the employer exercises that power, the sub-contractor will be entitled to recover any loss and expense which he incurs as a result.

(iii) Suspension by the sub-contractor

Although a valid exercise of the sub-contractor's right to suspend in the event of non-payment is a Relevant Event expressly entitling the sub-contractor to an extension of time, it is not included in the list of Relevant Matters entitling the sub-contractor to disruption costs. However, it is fairly clear that such costs are recoverable from the main contractor, on the grounds that the effective cause of the disruption was the main contractor's default, namely failing to pay the sub-contractor. Works Contract/2 is the only standard sub-contract which contains an express term to that effect (in clause 2.3.2), but it is suggested that the position is the same under the other forms.

Common law claims

All of the standard forms of sub-contract make it clear that the express provisions they contain entitling one party to claim loss and expense from the other are without prejudice to any other rights or remedies which the parties possess, which means that a claim for damages at common law will always be an option.

However, in order to establish such a claim the sub-contractor must be able to prove that the main contractor has acted in breach of an express or implied term of the sub-contract, which means that permitted actions such as ordering variations cannot give rise to common law claims.

Common law claims for damages due to delay and disruption are generally based on implied rather than express terms. Identifying the terms which will be implied into a standard form of sub-contract in respect of regular progress is no easy matter, as the courts have usually discussed this issue in relation to main contracts, where the considerations are somewhat different.

For example, in *London Borough of Merton v. Stanley Hugh Leach* (1985) 32 BLR 51, it was held that a term should be implied into the 1963 edition of the JCT Standard Form that the employer would not hinder or prevent the contractor from carrying out its obligations in accordance with the terms of the contract or from executing the works in a regular and orderly manner. Although a similar term is likely to be implied into sub-contracts, it will be modified by the express terms of the sub-contract, and all of the standard sub-contracts require the sub-contractor to carry out the sub-contract works reasonably in accordance with the progress of the main contract works. This means that the sub-contractor cannot expect complete continuity of working.

In addition, there may be indications in the sub-contract documents as to the order in which work is to be carried out or a programme may be agreed between the sub-contractor and main contractor. If so, it is those documents which will set the standard against which the main contractor's conduct should be judged, rather than any implied term.

Sub-contractors should therefore beware of obligations such as that imposed in *Martin Grant* v. *Sir Lindsay Parkinson*, which required the sub-contractor to 'proceed with any portion . . . of the work at such time . . . as the contractor should require having regard to the requirements of the contractor in reference to the progress or conditions of the main works'. In that case it was held that, by agreeing to such a term, the sub-contractor had lost the right to claim disruption costs from the main contractor, as the express term was sufficient to negate any implied term that the main contractor would do nothing to prevent the sub-contractor from completing his works.

The rules governing the quantification of common law damages are similar to those which apply to the calculation of loss and expense which are set out below; the material differences are summarised at the end of the section headed 'Quantification'.

There are, however, two important differences between a claim for loss and expense under the contract and a claim for damages at common law, namely:

(i)   the contractual procedures which apply to the recovery of loss and expense do not apply to the recovery of damages, so if the sub-contractor has failed to issue the appropriate notices under the sub-contract this will not affect his common law rights; *and*

(ii)   neither the architect nor the quantity surveyor has an inherent power to ascertain or negotiate damages claimed at common law, although they do have such powers in relation to loss and expense.

Notice requirements and procedures

If the sub-contractor wishes to make a claim under the sub-contract for loss and expense, he must submit the relevant notices and follow the correct procedure. Failure to do so will have much more serious consequences than failing to issue notices of delay, because the sub-contracts provide that the sub-contractor's right to recover is subject to the following provisos:

(i)   the sub-contractor has made a written application to the contractor for loss and expense as soon as it has or should have become reasonably apparent that the regular progress of his work has been or is likely to be affected; *and*

(ii) the sub-contractor has supplied information to the main contractor to demonstrate that his progress has or is likely to be materially affected (under NSC/C and Works Contract/2 the obligation is to submit such information as is requested by the architect through the main contractor);

(iii) the sub-contractor has supplied such details of the loss and expense as are reasonably requested by the main contractor (once again, the request will originate with the architect under NSC/C and Works Contract/2).

Failure to comply with these requirements is likely to mean that the sub-contractor has lost his right to recover loss and expense under the sub-contract, although there are no decided cases on the point. He will be forced to make a claim for damages at common law if he wishes to recover disruption costs but, as stated above, that will be of no assistance if the disruption was caused by an action permitted by the contract, such as the ordering of variations.

It is important to appreciate that the procedural requirements summarised above do not amount to a contractual requirement to submit a formal claim document, although in practice this may be the best way to present the information.

Quantification of direct loss and/or expense

The precise term used by the standard forms of sub-contract to describe the sums recoverable by the sub-contractor in the event that his regular progress is materially affected by a 'Relevant Matter' or by the act, omission or default of the main contractor is 'direct loss and/or expense'. In *F.G. Minter v. Welsh Health Technical Services Organisation* (1980) 13 BLR 1 the Court of Appeal held that this was such loss and expense which arose naturally and in the ordinary course of things – in other words, it should be quantified in accordance with the same principles as those laid down in the first rule in *Hadley v. Baxendale* for the quantification of damages for breach of contract.

Two important principles follow from this basic rule. Firstly, the loss and expense must equate to the actual costs incurred by the sub-contractor. This is where an excessively technical approach to the preparation of claims falls down, as the use of sophisticated formulae to calculate overheads often produces figures which are way in excess of any additional costs the sub-contractor has actually incurred. Similarly, a simple calculation based on the preliminaries as priced in the bills of quantities will rarely produce a figure which relates to actual cost (see below under the heading 'Increased Preliminaries' for further details).

Secondly, the function of damages, and by analogy of direct loss and expense, is to put the sub-contractor in the same, but no better, position than he would have been in had the contract been properly performed. This means that if, for example, the sub-contractor is claiming loss and expense due to variations, he should take into account when calculating the overhead and loss of profit elements of his claim that the valuation of the varied work itself should have provided him with an element of overhead and profit recovery.

Bearing in mind these general principles, the following 'heads of claim' will generally make up the direct loss and/or expense which a sub-contractor is entitled to recover:

Increased preliminaries

Preliminaries is the term used to describe costs which the contractor must incur in carrying out the work which do not form part of the work itself and which, in a traditional bill of quantities, are collected together in the first section of the bill. For a main contractor, they will include site supervision costs, insurance premiums, plant and tools and site establishment costs such as huts, water and electricity. A sub-contractor's preliminaries are likely to be more restricted because he will usually have use of the main contractor's site set up free of charge, although he will, of course, still incur costs in relation to plant and tools, supervision and insurance.

Because some preliminary costs are time-related, if there is a delay in completion they will be increased, and the additional costs so incurred will be recoverable as direct loss and expense.

A popular approach to quantifying a claim for extended preliminaries is to take the sum included for preliminaries in the bill of quantities or other priced document and to convert it into a weekly charge by reference to the original contract period. That weekly rate is then used to calculate the additional preliminaries costs for the period of the delay. Although a claim calculated on that basis will be accepted by many quantity surveyors, it is suggested that such an approach is fundamentally flawed, because it ignores the basic rule explained above that loss and expense must equate to actual cost.

There are several reasons why a weekly rate will not represent actual costs, for example:

- the sum allocated against the preliminaries item in the priced document by the sub-contractor can only be an estimate of the actual cost, and in practice is often an arbitrary figure;
- certain costs which are included in the 'preliminaries' figure in the priced document will be in respect of one-off items which are not time related and will therefore be unaffected by any delay or disruption;

- contractors often choose to spread some of the cost of items listed in the preliminaries section over the rates for the work itself, and therefore in a claim for variations, for example, they will be reimbursed for some preliminary costs in any payment based on the contract rates.

A better, although admittedly more time consuming, approach to claiming increased preliminary costs is to deal with each item separately, and to demonstrate the actual additional costs which have been incurred due to the delay or disruption. This will be reasonably straightforward when claiming items such as additional premiums charged for performance bonds being maintained for the period of delay, but can be more complex when calculating the costs which should be claimed for items such as plant and tools.

If the sub-contractor had hired the plant and the period of hire had been extended because of the delay or disruption, the sub-contractor will usually be entitled to recover the additional hire charges. However, if the plant was lying idle for part of the extended period, the sub-contractor will have to prove that it was not reasonable for him to have deployed the plant on another site or to have terminated the hire agreement and re-hired the plant when it was required on site.

If the sub-contractor had been using his own plant and equipment, the courts have held in *Bernard Sunley and Co. Ltd* v. *Cunard White Star* [1940] 2 All ER 97 that his loss and expense should be assessed on the basis of depreciation, maintenance and operator's costs and loss of interest on the capital investment in the plant. If the sub-contractor wishes to charge any greater sum, he will have to prove that, but for the delay or disruption, he would have been able either to hire out the plant elsewhere or to use it on another job to earn a contribution to his general overheads and profit.

Another major element of most claims which usually appears under the heading of additional preliminaries is supervision costs. Delay and disruption create logistical difficulties for both main contractors and sub-contractors and if, as sometimes happens, a sub-contractor allocates a senior manager to the site because of these problems, the costs he incurs as a result should be recoverable. However, any sub-contractor attempting to claim such costs must be able to prove that the allocation of a senior individual was reasonable and to produce evidence of the amount of time he spent on the project in the form of time sheets and diaries.

Overheads

There are two types of overheads which may form part of direct loss and expense: those which relate specifically to the delayed contract, and the general overheads of running the sub-contractor's business.

If head office staff spend more time on the administrative aspects of a particular contract than they would otherwise have done, for instance because of the need to organise overtime working or periods of lay-off, the additional costs incurred by the sub-contractor as a result will be recoverable, provided they can be proved. In cases of serious disruption, it may even be necessary for a major sub-contractor to take on additional staff to deal with the situation, and provided that was a reasonable step to take, the sub-contractor will have a valid claim for the costs of doing so. There should be no overlap here with costs claimed as additional preliminaries, as they are concerned with site based staff.

Claims for such specific additional overheads are relatively rare, but where they can be established they tend to be reasonably uncontentious because both the fact that the sub-contractor has incurred loss and expense and its amount can be demonstrated. The recovery of general head office overheads is, however, much more problematic.

Head office overheads cover all of the costs which the sub-contractor incurs in running his head office such as clerical staff, premises costs, stationery and office equipment. Most sub-contractors will include an element of head office overheads in any claim for loss and expense as a matter of course, but in fact they will need to establish firstly that they are entitled in principle to recover such overheads, and secondly that the method they have used to calculate the relevant figure is a reasonable one.

The legal basis of a claim for head office overheads was neatly summarised by Sir William Stabb QC in *J.F. Finnegan Ltd* v. *Sheffield City Council* (1988) 43 BLR 124 when he said:

'It is generally accepted that, in principle, a contractor who is delayed in completing a contract due to the default of his employer, may properly have a claim for head office overheads during the period of delay, on the basis that the workforce, but for the delay, might have had the opportunity of being employed on another contract which would have had the effect of funding the overheads during the overrun period. The principle was approved in the Canadian case of *Shore & Howitz Construction Co. Ltd* v. *Franki of Canada* (1967) SCR 589, and was also applied by Mr Recorder Percival QC in the unreported case of *Whittal Builders Company Limited* v. *Chester-le-Street District Council.'*

However, the language used by Sir William Stabb in this extract casts some doubt upon the ability of contractors and sub-contractors to claim head office overheads if there was no other work available to them during the period of delay which, but for the delay, they would have had a realistic chance of obtaining. Several leading commentators have expressed the view that head office overheads are not recoverable in such circumstances – notably the learned editor of *Keating* (5th Edition at p. 211) – which is a significant restriction during an economic recession.

If, however, the sub-contractor can negotiate the first hurdle of proving that he has a valid claim in principle for the recovery of head office overheads, he will then have to find a reasonable basis on which they can be calculated. One of the most popular methods is by using a formula such as that put forward by the learned editor of *Hudson on Building Contracts*, known somewhat unoriginally as the Hudson formula (see p. 599, 10th Edition). It deals with the calculation of both the overheads and profit elements of a claim for loss and expense by the following equation:

$$\frac{\text{HO/Profit \%age}}{100} \times \frac{\text{Contract sum}}{\text{Contract period}} \times \text{Period of delay (in weeks)}$$

where HO/Profit %age is the allowance made by the sub-contractor for overheads and profit in his original tender for the work.

Although there is some support for the use of the Hudson formula in *J.F. Finnegan* v. *Sheffield City Council* it has been heavily criticised in many of the leading textbooks on claims, notably *Building Contract Claims* by Vincent Powell-Smith and John Sims (2nd edition BSP Professional Books at pp. 130–132). Because it is based on the allowance made by the contractor in his tender it can certainly be attacked on the basis that it does not equate to the sub-contractor's actual loss, and it can produce some extraordinarily high figures which in themselves undermine its credibility.

It is suggested that if any formula is to be adopted, the most realistic is that put forward in *Emden's Building Contracts and Practice 8th Edition*, vol. 2 p. N/46, which is as follows:

$$\frac{h}{100} \times \frac{c}{cp} \times pd$$

h = head office percentage calculated by dividing the total overhead cost and profit of the sub-contractor's organisation as a whole by the total turnover

c = the contract sum

cp = the contract period

pd = the period of delay

with both the contract period and the period of delay being calculated in the same units, for example weeks.

The main advantage which this has over Hudson's formula is that it is based on the sub-contractor's overhead and profit recovery on his whole business rather than just on the delayed contract.

However, all formulae are susceptible to attack on the basis that, in reality, most major sub-contractors' organisations are sufficiently large and flexible to cope with delay and disruption on one contract without their ability to secure other work being adversely affected, so that they actually suffer no loss at all under this head of claim. Smaller sub-contractors are much more likely to be prejudiced, but as they may be less sophisticated, they may fail to present what is, for them, a perfectly valid head of claim in an acceptable way.

Profit

In principle, loss of profit is an acceptable head of claim, but the sub-contractor must be able to prove that the delay or disruption prevented him from earning profit on other contracts in the normal course of his business. As with overheads, this means that claims for loss of profit are much more difficult to sustain in times of recession.

Further, if the sub-contractor wishes to recover an unusually high profit margin on a contract he has lost because of the overrun or disruption, under the second rule in *Hadley* v. *Baxendale* he will have to prove that the main contractor was aware when he entered into the sub-contract that such a loss was likely in the event of a delay.

Disruption and loss of productivity

It is generally accepted that delay and disruption lead to a loss of productivity on the site, but quantifying and proving the costs incurred is an extremely difficult task. As a result, this head of claim is often the subject of an estimated allowance; the validity or otherwise of such a tactic is discussed below under the heading 'The global approach'.

This is an area where the separation of the concepts of delay and disruption is particularly important: the sub-contractor is often able to complete his work within the original sub-contract period, but the sequence of his work is so disrupted that his costs are far higher than they should have been. In such circumstances, the sub-contractor will have a valid claim for disruption costs, but he will have to produce evidence to show that his actual method of working was less productive than the method he had planned to adopt. This can be done by means of histograms and other charts showing both the actual and planned deployment of labour and other resources. Although they are time-consuming to prepare, such charts will usually result in serious consideration being given to this element of the sub-contractor's claim.

The difficulties of quantification are compounded for sub-contractors by the fact that, even if everything runs smoothly on the project, they are not entirely in control of the progress of their work because of their general obligation to work reasonably in accordance with the main contract works. On the other hand, it can be argued that the comparison between antici-pated and actual progress referred to above is more straightforward for a sub-contractor because his activities will often be the same on every contract he undertakes, with the result that it will be easier for him to establish what would have amounted to 'regular progress' on the disrupted contract.

Increased costs due to inflation

Increases in labour and materials costs which the sub-contractor has suffered because of the delay or disruption can be recovered as direct loss and expense. If, for example, the sub-contractor's start on site is delayed by a month, assuming the rest of his progress is uninterrupted, each month he will be able to claim the additional costs he has incurred because he is carrying out work a month later than he had anticipated when he tendered.

This is one of the rare instances when the use of a formula can be justified: the NEDO formula for calculating increased costs is generally accepted as a reasonable basis for quantifying such costs.

If the sub-contract was let on a fluctuating price basis (which is exceedingly rare in the current climate of low inflation) the sub-contractor would have to take into account the price increases he recovered under the fluctuations clauses when calculating increases in cost claimed as loss and expense in order to avoid any double recovery.

Financing charges

The case of *F.G. Minter* v. *Welsh Health Technical Services Organisation* (1980) established that financing charges were a valid element of a claim for direct loss and expense under the JCT standard forms. In the Court of Appeal, Lord Justice Stephenson said:

> '[in] the building industry the "cash flow" is vital to the contractor and delay in paying him for the work he does naturally results in the ordinary course of things in his being short of working capital, having to borrow capital to pay wages and hire charges and locking up in plant, labour and materials capital which he would have invested elsewhere. The loss of interest which he has to pay on the capital he is forced to borrow and on the capital which he is not free to invest would be recoverable from the Employer's breach of contract within the first rule in *Hadley* v. *Baxendale* (1854) without resorting to the second, and would accordingly be a direct loss, if an authorised variation of the works, or the regular progress of the works having been materially affected by any event specified . . . has involved the contractor in that loss.'

The financing charges which the contractor is therefore entitled to recover will be either the interest he has had to pay on the additional capital needed to finance the work during the period of the delay or disruption or, if he is in credit, the amount of interest he would have earned on the capital if he had been free to invest it elsewhere.

Once a claim to financing charges has been established in principle, there are two subsidiary issues which must be considered. The first relates to the applications for financing charges which must be issued by the contractor or sub-contractor. In *Minter*, where the contractor's claim was under JCT 63, it was held that interest was recoverable only for the period between the loss and expense being incurred and the date of the contractor's written application for its reimbursement. This is because in JCT 63, the drafting of the clause entitling the contractor to recover loss and expense refers only to losses which have already been incurred; it does not relate to future losses. Accordingly, in any claim for interest based on JCT 63, the contractor is bound to make regular applications for the payment of interest.

This requirement was modified slightly by the Court of Appeal in *Rees & Kirby Ltd* v. *Swansea City Council* (1985) 30 BLR 1, when they stated that the contractor's application for loss and expense need not be in any particular form, but that it must be clear that it included an element of loss and expense because the contractor had been stood out of his money.

JCT 80 and the other main forms based upon it (which includes all of those within the scope of this book) provide for the recovery of continuing or future losses, and therefore a single notice that financing charges are and will continue to be incurred should be sufficient under those contracts. NSC/C, Works Contract/2 and NAM/SC are drafted on a similar basis to JCT 80, but DOM/1 (and therefore DOM/2) and IN/SC refer to 'direct loss and expense thereby **caused**', and so it could be argued that under those forms, regular applications are necessary.

The second issue is whether the interest should be calculated on a simple or a compound basis. This was addressed in *Rees & Kirby* which established that, where the contractor was being forced to pay compound interest because he was being stood out of his money, he would be entitled to recover financing charges calculated on the same basis as part of his loss and expense.

Costs of claim preparation

It is generally accepted that main contractors are not entitled to recover any costs they incur in the course of preparing a claim for loss and expense, regardless of whether they have prepared the claim themselves or appointed a claims consultant to do it for them. This is because, under the contract, the quantity surveyor is obliged to ascertain any loss and expense, so, in theory, the contractor's only role should be to provide him with such information as he requests to enable him to carry out that task.

If, however, the dispute gets as far as arbitration or the courts, any sums paid to a claims consultant for the preparation of the contractor's case for those purposes may be awarded to the contractor by the court or arbitrator, as was done in the case of *James Longley & Co. Ltd* v. *South West Regional Health Authority* (1984) 25 BLR 56.

It is suggested that sub-contractors will be in a similar position to main contractors in relation to this head of claim, although domestic sub-contracts such as DOM/1 provide for loss and expense to be 'agreed' between the parties rather than ascertained by either the quantity surveyor or the main contractor. This may give the sub-contractor more scope to argue that, in order to reach such an agreement, it was essential for him to prepare a detailed claim document, but there are no decided cases on the point and in their absence claims for such costs remain highly speculative.

Some support may be gleaned from the case of *Tate & Lyle Food and Distribution Co. Ltd* v. *Greater London Council* [1983] 2 AC 509, in which it was held that 'the expenditure of managerial time in remedying an actual wrong done to a trading concern can properly form the subject matter of a head of claim'. However, the judge in that case refused to award any costs under that head of claim because of a lack of evidence, and expressly declined simply to add a percentage of the other sums he had awarded. It should also be borne in mind that the costs of additional management time may have already been recovered as overheads.

Global claims

Given the length and complexity of this chapter, it is not surprising that contractors sometimes try to avoid producing a detailed justification of their case and rely on a 'global' or 'rolled up' claim. These terms are used to describe claims which simply list the causes of delay and disruption and state that an additional (usually substantial) sum of money as loss and expense has been incurred as a result, without linking cause and effect or quantifying the loss in detail. They are certainly quicker and easier to produce than a properly quantified and argued claim, but they will be sustainable only in very limited circumstances.

Many contractors were encouraged to put forward global claims by the decision in *J. Crosby & Sons Ltd* v. *Portland Urban District Council* (1967) 5 BLR 121. In that case, the employer was challenging an award by an arbitrator of a lump sum in respect of loss and expense caused by a 31 week delay on the grounds that the arbitrator could not award loss and expense without attributing specific amounts to each head of claim. The judge upheld the arbitrator's finding, and said:

'I can see no reason why he [the arbitrator] should not recognise the realities of the situation and make individual awards in respect of those parts of individual items of claim which can be dealt with in isolation and a supplementary award in respect of these claims as a composite whole.'

Unfortunately this perfectly logical and, it is suggested, correct statement of the law has led some contractors and their lawyers to put forward totally unparticularised claims which make no attempt to link cause and effect. The most notable example of this to come before the courts is *Wharf Properties* v. *Eric Cumine Associates* (1991) 52 BLR 1, which concerned an action by an employer against his architect for the costs of delay and disruption to a project. The statement of claim (the pleading setting out the plaintiff employer's case) did not link the six specific periods of delay which had occurred to the architect's breaches of duty, on the grounds of the complexity of the project and the large number of delaying factors which made it impossible to identify individual delays.

The Privy Council totally rejected that argument. Although they accepted the proposition put forward by judge in *Crosby* v. *Portland* quoted above regarding quantification, this did not excuse a plaintiff from pleading his case in such a way as to alert the defendant as to the case which would be made against him at trial. The pleading was therefore struck out, because it made no attempt to show a connection between the wrong alleged and the consequent delay, and so provided 'no agenda for the trial'.

Both *Wharf* v. *Cumine* and subsequent cases such as *ICI plc* v. *Bovis Construction Ltd* (1992) 32 CON LR 90 have made it absolutely clear that, although a global may be permissible in the sense that a composite figure can be claimed where it is impossible or impractical to trace the loss back to a specific event, the claimant must show a connection between the wrong complained of and the delay or disruption which caused that loss.

Quantification of damages at common law

The principles set out above apply equally to the quantification of a claim at common law for damages for breach of contract, with the important exception of interest and financing charges.

This exception arises because the cases described above which establish the contractor's right to financing charges were specifically concerned with their recovery as part of the loss and expense payable under the JCT standard forms. If a contractor or sub-contractor is claiming damages for breach of contract he will be unable to rely on *Minter* v. *WHTSO* etc., and will have to overcome the traditional reluctance of the courts to award interest.

At common law, the general rule is that if a debt is paid late there will be no right to interest unless there is either an express power to charge interest in the relevant contract or legal proceedings have been commenced for its recovery. This has been confirmed in numerous cases, and was recently reiterated by the House of Lords in *President of India* v. *La Pintada Cia Navigacion SA* [1985] 5 AC 104.

It may nevertheless be possible to recover interest as part of a common law claim if it can be established that the interest constitutes an item of special damage under the second rule in *Hadley* v. *Baxendale*. The sub-contractor will need to prove that, as a direct result of the contractor's breach of contract, he has incurred interest or financing charges, and that the contractor was aware before he entered into the sub-contract of the circumstances which caused that loss. This principle was confirmed in *Holbeach Plant Hire Ltd* v. *Anglian Water Authority* (1988), but the court stressed the need both for proof and for the case to be pleaded in the correct way.

If legal proceedings have been started, the courts have a statutory power to award interest under section 35A of the Supreme Court Act 1981, and arbitrators have a similar power under section 19A of the Arbitration Act 1950.

Negotiation of claims

Virtually all claims, however carefully and honestly compiled, will be subject to some negotiation. There are several basic rules which should be borne in mind by all parties during this process.

Firstly, a properly prepared claim based on sound legal principles should be treated seriously. If it is ignored, the claimant can pursue his legal rights in arbitration or the courts (for a full commentary on the different procedures available, see Chapter 12).

Secondly, requests for further information may be appropriate, but should not be relied on as a means of keeping the claim at bay indefinitely. If the claimant conscientiously responds to requests for further information (or gives legitimate reasons as to why it is unnecessary or unavailable), there will eventually be no further requests of this type which can properly be made.

Thirdly, it is in the interests of everyone for those involved in the negotiation of claims to have the authority to reach a settlement. Endless reference back to others can not only be frustrating, but can lead to the withdrawal of settlement offers or the unnecessary escalation of the dispute.

Finally, when an agreement is reached, it should be reduced to writing as quickly as possible. If it is achieved at a meeting, a document should be drawn up there and then and signed by all those present.

10 Damage and Insurance

There are innumerable risks involved in the construction of a building: the risk of damage to the works themselves or to materials on site, the risk of death or physical injury to the contractor's and sub-contractors' employees or to visitors or passers-by, and the risk of damage to any adjoining property. The JCT standard forms of contract and their related sub-contracts deal with all of these risks and, in some cases, require the person who bears the risk to insure against it.

Taking JCT 80 as an example, this chapter describes how the various risks are allocated and what insurances the parties are required to carry. It also deals with the related issue of protection of the sub-contract works.

Because the contractual provisions dealing with these issues are treated as a model by the Tribunal, and are therefore incorporated in all of its standard forms with only very minor amendments, the other main and sub-contracts are not commented upon in detail. Any material differences have been highlighted in the section headed 'Other JCT Forms'.

Although none of the JCT forms imposes an obligation on the contractor or his sub-contractors to carry professional indemnity insurance, if they are involved in design such insurance is often required by clients. Similarly, none of the standard forms refers to performance bonds or parent company guarantees, but many clients demand them. Both of these subjects are therefore discussed at the end of this chapter.

Damage to the Works – JCT 80

In order to understand the rules in the sub-contracts as to the allocation of risk and insurance obligations relating to damage to the sub-contract works, it is necessary to appreciate the arrangements which apply at main contract level.

Clause 22 of JCT 80 provides for three alternatives in this regard:

Clause 22A: new buildings – insurance by contractor
Clause 22B: new buildings – insurance by employer
Clause 22C: existing buildings – insurance by employer.

If the main contract is for the construction of a new building, the employer must decide before he invites tenders for the main contract works whether he wishes the contractor to take out the insurance or whether he prefers to make his own arrangements. If the contract is for work in or an extension of an existing building he has no choice: clause 22C will automatically apply and the employer will be obliged to insure both the works and the existing structure and its contents.

Each of the three alternatives in the main contract is described below.

Clause 22A: new buildings – insurance by contractor

Where Clause 22A applies to the main contract, the contractor must take out and maintain all risks insurance for the full reinstatement value of the works (which includes all work carried out by sub-contractors, whether nominated or domestic).

The policy must be in the joint names of the employer and the contractor, which means that the employer will not need to take out his own insurance against damage to the works or to materials on site caused by his own negligence. Any such damage will be covered by the joint names policy and, because the employer is named as an insured on the policy, the insurers will not be able to exercise any subrogation rights against the employer. Subrogation rights is the technical term for rights the contractor may otherwise have had in respect of any negligence by the employer which pass to the insurers when they indemnify the contractor against the consequences of that negligence.

The principle behind requiring the policy to be in joint names is an extremely sensible one: if there is no need to establish the precise cause of the damage, the only payments made under the policy will be in respect of reinstatement, that is rectification work. The substantial lawyers' fees which would otherwise be incurred in establishing who, if anyone, was negligent, are avoided, and the total additional cost to the building process is kept to a minimum.

The minimum cover which must be provided by the all risks insurance is defined in clause 22 of JCT 80. The policy must provide cover against any physical loss or damage to work executed or to goods and materials on site. Thus damage due to virtually all causes, including what the insurance industry describes as the 'perils' of fire, explosion, flood, theft, vandalism and accidental damage will all be covered.

Certain limited exclusions are permitted, namely:

- property which is defective due to wear and tear, obsolescence, deterioration, rust or mildew;
- work or materials damaged through a defect in their own design or workmanship, or loss or damage to any other work which relied on the defective work or materials for support or stability;
- loss or damage caused by or arising from:
  - war, invasion, etc.;
  - disappearance or shortage revealed by an inventory or not traceable to an identifiable event;

- an Excepted Risk (which is defined in clause 1.3 of JCT 80 as the nuclear risks and the risks of damage due to aircraft travelling at sonic or supersonic speeds).

If the contract is carried out in Northern Ireland, there is no requirement to insure against the risk of damage due to civil commotion or terrorist acts by outlawed organizations such as the IRA. This is because there is a government compensation scheme which deals with damage caused by terrorists in Northern Ireland.

The increase in terrorist activity on mainland Britain has recently resulted in new insurance arrangements being established in the rest of the UK for fire and explosion due to terrorism, which involve participation by the government as ultimate reinsurer.

Only £100 000 worth of cover against such damage is now automatically included in all risks insurance policies. If more cover is required, the insured will have to pay an additional premium which will vary according to the value of the project, the location of the site, and whether the contract is likely to be a terrorist target. In spite of these special arrangements, cover for damage to the works due to terrorism remains a standard element of the all risks cover as it is defined in JCT 80, and so contractors who are required to insure under 22A will have to take into account the additional premium charged for terrorist cover when they tender for the work.

The all risks policy must remain in place until practical completion of the main contract works, or until the determination of the contractor's employment, whichever occurs first.

Many main contractors will not take out a one-off policy for each contract, but will have an annual all risks policy which covers all of the projects on which they are working to which clause 22A (or its equivalent in other standard forms) applies. This is recognised in and permitted by the contract, although the contractor must inform the employer of the renewal date of his annual policy so that he can check that the policy is being maintained and the premium paid.

If a one-off policy has been arranged by the contractor to discharge his obligations under clause 22A it must be sent, together with a receipt for the premium, to the architect who deposits it with the employer.

If the contractor fails to comply with his obligation to take out insurance in accordance with Clause 22A, the employer may take out an appropriate policy and recover the cost of doing so from the contractor as a debt.

If the works are damaged by one of the risks insured by the all risks policy, the contractor must repair the damage and proceed with the carrying out and completion of the works. This will be the case regardless of whether there is only minor damage or the whole building has burnt down the day before practical completion.

The contractor authorises the payment of the insurance monies due under the all risks policy to the employer, who releases them to the contractor in instalments as the reinstatement is carried out and interim certificates are issued. The contractor is not entitled to recover any more than the insurance monies for the repair or restoration work, and thus he bears any excess on his insurance policy.

Clause 22B: new buildings – insurance by employer

The arrangements where clause 22B applies are similar to those described for clause 22A, save that it is of course the employer and not the contractor who is obliged to take out the all risks policy in joint names.

One significant difference is that if damage occurs which is covered by the all risks policy, the contractor must rectify it, but is paid for doing so as if the work had been required by a variation instruction. This means that the contractor does not bear any excess applicable to the all risks policy. He will also be able to claim any loss and expense which he incurs because of the damage to the works, including any increase in the cost of completing any undamaged portions of the works which may arise if a fire, for example, delays the completion of the whole project and the cost of materials therefore increases.

Some employers, particularly in the public sector, do not buy insurance, but prefer to carry the risk themselves. Although JCT 80 itself does not cater for this, when the Tribunal issued the current insurance clauses in 1986, they published Practice Note 22, a Practice Note and Guide to the clauses which includes several model clauses which may be used in place of clause 22B. These clauses allow the employer to take either the 'risk' or the 'sole risk' of loss or damage to the works, but do not require him to insure against it.

The distinction between 'risk' and 'sole risk' is an extremely important one. Where the employer takes the sole risk of loss or damage to the works, he will be obliged to pay for their reinstatement regardless of whether the damage was caused by the contractor's (or any sub-contractor's) negligence. If, however, the employer takes only the 'risk' of damage, he will bear the reinstatement costs only where the damage is not caused by the negligence of the contractor or any of his sub-contractors. This is contrary to the philosophy of the standard JCT 80 provisions, which is to avoid expensive legal argument as to the cause of the problem and simply to rectify it and get on with the work. The model clauses in Practice Note 22 have, however, proved very popular with local authority employers.

Clause 22C: existing buildings – insurance by employer

Clause 22C deals separately with the insurance of the works and the obligation to insure the existing structure and its contents.

The works must be insured on an all risks basis in exactly the same way as where they are new building works being insured by the employer under clause 22B. Model clauses enabling the employer to take the risk or sole risk of damage to the works without having to insure are also set out in Practice Note 22.

The employer must insure the existing building and its contents which he either owns or for which he is responsible against loss or damage due to what are described as the 'Specified Perils'.

The Specified Perils are:

- fire
- lightning
- explosion
- bursting or overflowing of water tanks, apparatus or pipes
- earthquake
- aircraft and other aerial devices or articles dropped from them (excluding damage caused by sonic or supersonic booms)
- riot and civil commotion.

The policy must be in the joint names of the contractor and the employer.

The effect of this more limited cover is that if the existing building is damaged by a Specified Peril, for example a fire, the contractor will not be responsible for any costs the employer may incur as a result, even if the fire was caused by the contractor's negligence. If, however, the damage was caused by something other than a Specified Peril, for example a JCB had been driven into part of the existing building (a risk described rather quaintly by the insurance industry as 'impact'), the contractor would be responsible for any costs incurred by the employer assuming, of course, that the JCB driver was negligent. The contractor's liability in those circumstances would normally be covered by his public liability policy.

If work being carried out in an existing building is lost or damaged, either the employer or the contractor may determine the contractor's employment if, in the words of the contract, 'it is just and equitable to do so'. This is most likely to occur where there has been significant damage to the works which also affects the existing building.

Damage to the nominated sub-contract works – NSC/C

If the principle of a single insurance policy covering all damage to the works, regardless of any negligence, was applied at sub-contract level, all sub-contractors would be named as insured under the all risks policy taken out under the main contract, and insurers would not therefore have any subrogation rights against them. Unfortunately for sub-contractors, when the current insurance clauses were discussed with representatives of the UK insurance market in 1986, they refused to extend the full cover of the main contract all risks policy to sub-contractors.

However, nominated sub-contractors are given the benefit of the majority of the main contract all risks cover by the standard nominated sub-contract conditions NSC/C.

Damage caused by the Specified Perils

Clause 6 of NSC/C contains three alternative provisions, clauses 6A, 6B and 6C, which correspond to clauses 22A, 22B and 22C of JCT 80. Where clause 22A applies to JCT 80, clause 6A applies to NSC/C, where clause 22B applies clause 6B applies to NSC/C and where clause 22C applies to the main contract clause 6C applies to the sub-contract. All three have similar provisions regarding damage to the sub-contract works.

Under clause 6A, 6B or 6C, whichever is applicable, whoever is responsible under the main contract for taking out all risks insurance of the main contract works must ensure that, **in respect of damage caused by the Specified Perils only**, the nominated sub-contractor is either recognised as an insured under the policy or the insurers waive any rights of subrogation they may have against the sub-contractor. That recognition or waiver must continue until either the date the sub-contractor's employment is determined or the date of practical completion of the sub-contract works, whichever occurs first (described in NSC/C as the 'Terminal Dates').

The effect of this is that the sub-contractor will not be responsible for damage to his work due to a Specified Peril even if it was caused by his own negligence.

These provisions of NSC/C are often ignored or misunderstood by main contractors, who regularly set-off monies from sub-contractors if the sub-contract works have been damaged by a Specified Peril caused by the sub-contractor's negligence. Such a set-off is a breach of contract by the main contractor, as the following example illustrates.

Suppose the nominated sub-contractor is carrying out mechanical and electrical work involving welding. Clause 22A applies to the main contract, so the contractor is insuring the works on an all risks basis. A welder employed by the sub-contractor negligently causes a fire.

Fire is a Specified Peril. Under clause 6.5 of NSC/C, the occurrence of loss or damage due to a Specified Peril must be ignored when calculating the sums paid to the sub-contractor, so the main contractor cannot refuse to pay for work already carried out by the sub-contractor which has now been damaged by the fire.

Clause 6A.2.1 provides that the sub-contractor is responsible for the cost of any restoration etc. of sub-contract work **except** to the extent such loss or damage is due to:

'one or more of the Specified Perils (whether or not caused by the negligence, breach of statutory duty, omission or default of the Sub-Contractor or any person for whom the Sub-Contractor is responsible)'.

Clause 6A.4 then goes on to state that, where the sub-contractor is not responsible for the cost of restoration etc. under clause 6A.2, it shall be valued as if it were a variation. The sub-contractor will also be entitled to an extension of time for any delay caused by the occurrence of the fire under clause 2.6.3 of NSC/C.

The insurers of the all risks policy taken out under the main contract, from whom the main contractor will recover any sums paid to the sub-contractor, will not have any subrogation rights against the negligent sub-contractor, because they will have either been required expressly to waive them, or the sub-contractor will be named on the policy in accordance with clause 6A.1.

The intention and effect of NSC/C is therefore absolutely clear: if damage is caused to the sub-contract works by a Specified Peril occurring before practical completion, the sub-contractor is fully protected by the joint names policy and should be paid in full for any rectification work.

Damage caused by other risks

There are several risks which could cause damage to the sub-contract works which are not included in the definition of Specified Perils under the JCT standard forms. Notable omissions include theft, vandalism, impact and subsidence.

Clauses 6A, 6B and 6C all deal with responsibility for damage by risks other than the Specified Perils in the same way.

Taking clause 6A as an example, clause 6A.2.2 of NSC/C provides that when materials and goods have been fully, finally and properly incorporated into the works before practical completion of the sub-contract works, the sub-contractor is responsible for their loss or damage only where it is caused by his own negligence or the negligence of someone for whom he is responsible (that is his own employees, agents and sub-sub-contractors).

Clause 6A.2.1 states that the sub-contractor is not responsible for loss or damage to either the sub-contract works or to goods and materials on site to the extent that it is due to the negligence or default of the main contractor, the employer, local authorities or statutory undertakers.

As clause 6A.2 makes it clear that, unless the sub-contractor is expressly relieved of the liability to rectify loss or damage, he must do so at his own cost, the net effect of these clauses is:

- the sub-contractor bears the risk of damage to work, goods and materials **before they are fully, finally and properly incorporated into the works**, unless the loss or damage is caused either by a Specified Peril or

by the negligence of the contractor, employer, a local authority or statutory undertaker;
* the main contractor bears the risk of damage to work, materials and goods **after they are fully, finally and properly incorporated into the works** unless the loss or damage is due to the negligence of the sub-contractor or someone for whom the sub-contractor is responsible.

Sub-contractors are not contractually obliged to take out insurance against the risks they bear in this regard, but they would obviously be well advised to do so.

Because of the way in which the risk of damage not caused by the Specified Perils is allocated, where a sub-contractor's goods and materials are stolen, or damaged by impact, subsidence or vandalism, there is often a dispute as to whether the relevant goods and materials have been fully, finally and properly incorporated into the works. There are no decided cases on the question of when goods and materials become fully, finally and properly incorporated, and NSC/C, like all the other standard forms of sub-contract, gives no guidance as to how the phrase should be interpreted.

Applying the standard legal rules on the construction of contracts, the words should be given their ordinary, natural meaning in the context in which they appear. Full, final and proper incorporation must, therefore, be capable of occurring before the sub-contract works are practically complete, as this is a possibility envisaged by the sub-contract in clause 6A.2.2 which opens with the words:

'Where during the progress of the Sub-Contract Works, sub-contract materials and goods have been fully, finally and properly incorporated into the Works before practical completion of the Sub-Contract Works.'

It is therefore fairly clear that something like a heating system does not need to be filled and tested before the goods and materials in it are fully, finally and properly incorporated. It is equally clear that where goods and materials are simply being stored on the site, and have not yet been worked upon, they are not fully, finally and properly incorporated. Between those two extremes, the question of when full, final and proper incorporation occurs is a matter of intense argument and speculation, which simply has to be resolved by agreement in each case.

It is sometimes argued that once work has been completed on a particular section of the work such as a room or a floor (regardless of any contractual provision for sectional completion), the good and materials in that section have been fully, finally and properly incorporated and therefore the risk of damage to them has passed to the main contractor. Although such an apportionment of risk would be logical, there is no guarantee that such an argument would persuade a main contractor or find favour with the courts.

In the event of theft or vandalism, sub-contractors often allege that the main contractor should pay for any loss or damage if he was contractually responsible for providing site security. Whether this is correct will depend on the precise circumstances of the case. The sub-contractor will have to prove that the main contractor's failure to provide proper security amounted to 'negligence, default or omission' on his part, which implies that he must have been in breach of a contractual obligation to the sub-contractor to provide a particular level of site security. The loss or damage must also be 'due to' that negligence, default or omission, so if it is clear that the lack of security did not contribute to or encourage the theft or vandalism, this argument will not assist the sub-contractor.

Damage to the domestic sub-contract works – DOM/1

The clauses in DOM/1 dealing with the risk and insurance of damage to the sub-contract works generally follow those in NSC/C, with two important exceptions.

Work in existing buildings

Where the main contract works are in or are an extension to an existing structure, clause 22C will apply to the JCT 80, and the employer will be required to insure the existing structure and its contents against loss or damage due to the Specified Perils. However, domestic sub-contractors will not be given the benefit of that policy by either being named on it as an insured or having subrogation rights against them waived, and therefore if they negligently damage the existing building or its contents, they will be responsible for the resulting costs.

On the face of it, this appears to be an enormous risk for the domestic sub-contractor, particularly if the sub-contract work is of relatively low value and the existing building is, for example, the National Gallery or a major hospital. However, the risk of such loss or damage should be covered by the domestic sub-contractor's public liability policy, which indemnifies him against the consequences of his own negligence.

Standard public liability policies exclude cover for damage to property in the custody or control of the insured, so if it is possible that the existing building is at any time in the custody or control of the domestic sub-contractor (for example, if it is empty because of the building work being carried out and only the sub-contractor's men are on site), it is essential for the sub-contractor to ensure that his policy is worded so as to ensure he has adequate cover. This will involve an amendment to, or the deletion of, the standard exclusion relating to property in the custody or control of the insured.

Terminal Dates

As explained in the commentary on NSC/C, the terminal dates mark the end of the period during which the sub-contractor has the protection of the main contract all risks policy. In the context of the nominated form, they are easy to define because practical completion of the nominated sub-contract works takes place on the issue of an architect's certificate.

A different approach is adopted in DOM/1 because of the different ways in which it can be proved under that form that practical completion of the domestic sub-contract works has been achieved, not all of which involve the issue of any sort of document. This is relevant because clause 22.3.2 of JCT 80 provides that domestic sub-contractors will be protected by the main contract joint names policy up to and including the date of issue of any certificate or other document which states that the domestic sub-contract works are practically complete.

DOM/1 provides for practical completion to occur on the date notified by the sub-contractor, or if the main contractor dissents from that date within fourteen days of receiving the sub-contractor's notice, on such date as may be agreed or determined by an arbitrator or, failing agreement and in the absence of arbitration, on the date of practical completion of the main contract works. If the date of practical completion is determined by agreement, the sub-contract does not provide for that agreement to be confirmed in writing, therefore the definition of the terminal dates in DOM/1 cannot simply refer to the date of practical completion of the sub-contract works.

Clause 6.1.2 of DOM/1 therefore defines the Terminal Dates relating to practical completion as:

'either the date of the written notice of the Sub-Contractor under clause 14.1 provided the Contractor does not dissent therefrom under clause 14.1 or, where the Contractor does so dissent, the date upon which the Contractor issues in writing to the Sub-Contractor a confirmation of the agreement under clause 14.2 or, failing such agreement, the date of issue of the certificate of Practical Completion of the Works under clause 17.1 of the Main Contract Conditions, whichever is applicable; or . . .'

Under this clause, the main contractor must confirm in writing any agreement reached with the sub-contractor as to the date of practical completion of the domestic sub-contract works in order to bring to an end the sub-contractor's protection under the joint names policy. A footnote has been included in DOM/1 pointing out that although written confirmation of any agreement is not required under clause 14.2 of DOM/1, it should be issued by the main contractor as soon as such agreement is reached, because of the impact of clause 6.1.2.

Protection of the sub-contract works

Sub-contract documents such as the specification often include a requirement that the sub-contractor must 'protect' his works until they are practically complete, or even until practical completion of the main contract works.

Clauses such as these are usually inserted by main contractors in an attempt to diminish their responsibility for damage to the sub-contract works. If the main contractor can claim that the sub-contractor is liable for any accidental damage caused by other trades working on the site on the grounds that it was due to the sub-contractor's failure properly to protect his work and is therefore damage caused by his own negligence, omission or default, he will not have to become embroiled in difficult arguments as to whether the relevant work has been fully, finally and properly incorporated.

Although all of the JCT standard forms require unfixed goods and materials on site to be 'adequately protected against weather and other casualties' they do not stipulate that the sub-contract works themselves should be protected in any way, and therefore an additional requirement to protect upsets the delicate balance of risk in the standard forms.

If no details are given as to the degree of protection required, such clauses can place an extremely onerous obligation on the sub-contractor. For example, a sub-contractor supplying a sophisticated controls system will find it difficult to protect his work from damage without going to a great deal of expense to provide special covers for his controls, and once he has left the site it will be impossible for him to ensure that his protection has not been disturbed or even removed by following trades.

Sub-contractors would therefore be well advised to try to negotiate the deletion of any clauses requiring them to protect their work, and if they cannot do so, to price the additional risk being imposed upon them.

Indemnity clauses and liability insurance

An indemnity is a promise by one party to make good a loss suffered by another. Both the nominated and domestic sub-contracts for use with JCT 80 contain broad indemnity clauses relating to death and personal injury and to property damage.

NSC/C and DOM/1 require the sub-contractor to indemnify the main contractor against any expense, liability, loss, claim or proceedings in respect of personal injury or death which arises out of the carrying out of the sub-contract works, except to the extent that it is due to the negligence of the main contractor (or his other sub-contractors), the employer or a local authority or statutory undertaker.

The sub-contractor's indemnity against damage to property is more limited, and relates only to property damage caused by his negligence, breach of statutory duty, omission or default. Further, the indemnity does not extend to loss or damage to the works or to site materials caused by a Specified Peril occurring before the Terminal Dates. This is to preserve the principle that such damage should be covered under the main contract all risks policy, without the insurers acquiring subrogation rights against any sub-contractor who negligently caused the Specified Peril.

The sub-contractor is contractually obliged to insure his liability under both indemnities, which he does by means of an employer's liability policy and a public liability policy (sometimes described as third party insurance).

The sub-contractor's employer's liability insurance must comply with the Employer's Liability (Compulsory Insurance) Act 1969, which requires all employers to insure their liability to their employees in respect of death or personal injury up to a limit of at least £2 000 000, although in practice the policy usually gives unlimited cover.

The minimum amount of public liability insurance which the sub-contractor must carry is stipulated in the appendix to the sub-contract.

Until 1987, JCT 80 required main contractors to ensure that all sub-contractors carried the same amount of public liability insurance which the main contractor was required to have under the main contract. On major contracts where high limits of indemnity were demanded from the main contractor this was simply impracticable for many smaller sub-contractors, so the contractual requirement was often ignored. This practice was recognised in Amendment 4 to JCT 80, which provided for the amount of public liability insurance to be carried by each sub-contractor to be a matter for agreement between the main contractor and the sub-contractor. The amount so agreed is then inserted in the appendix to the sub-contract.

Other JCT forms

As explained in the introduction to this chapter, the JCT treats the indemnity and insurance provisions of the standard forms as model clauses, and those in JCT 80 are therefore incorporated into the other standard forms with only minor amendments. There are, however, some material differences, which are summarised below.

Amount of public liability insurance required

The JCT Design and Build Form 1981 follows the approach now adopted in JCT 80 and permits the amount of public liability insurance to be carried by sub-contractors appointed under DOM/2 (or under a non-standard sub-contract if DOM/2 is not used) to be a matter for agreement between the sub-contractor and the main contractor.

However, IFC 84 still requires all sub-contractors, whether named or domestic, to carry the same amount of cover required of the main contractor under the main form.

The JCT Management Contract permits the employer to specify in the appendix to the Management Contract the level of public liability insurance which works contractors must carry: this may well be different to that required from the management contractor but is not entirely satisfactory, because it does not invite the employer to differentiate between the works contractor carrying out perhaps 20 per cent of the work on site and the works contractor who is carrying out only 2 per cent.

Risk of damage to existing structures and contents

Named sub-contractors under IFC 84 are equated with nominated rather than domestic sub-contractors for insurance purposes, and therefore if the work is being carried out in an existing building, they are given the benefit of the insurance policy taken out by the employer in respect of any damage caused by the Specified Perils to the existing structure and its contents.

Works contractors appointed under the JCT Management Contract are also treated in the same way as nominated sub-contractors in this regard.

Professional indemnity insurance

Professional indemnity insurance is a specialised type of insurance cover which was originally developed to indemnify professionals such as architects, solicitors and doctors against the damages they would be liable to pay if they were negligent in the performance of their professional duties.

Public liability insurance does not include cover for legal liability arising out of negligent design, and therefore if a sub-contractor wishes to insure against this risk, he will have to take out professional indemnity insurance.

Because many sub-contractors, particularly in the more specialist areas such as mechanical and electrical services, suspended ceilings and structural steelwork, almost invariably contribute to the design of their works, it is common for the employer to require them to carry this type of insurance.

A legal obligation to take out professional indemnity insurance may be imposed through the medium of a collateral warranty, which may require the sub-contractor to maintain the cover for up to fifteen years after practical completion of the works. The employer will want the insurance to remain in place for that length of time because all professional indemnity policies are written on a 'claims made' basis, which means that the insurers will be liable to pay only those claims which are made during the period of the policy, regardless of when the negligent act giving rise to the claim occurred. As the sub-contractor's legal liability in respect of any negligent design continues for up to fifteen years after the negligent act, the employer will want the insurance to remain in place for a similar length of time.

Like many other types of insurance, professional indemnity is available only on an annual basis, so it would be very rash of a sub-contractor to make such a commitment without some qualification as to the availability and price of such insurance in future years. A common amendment sought by sub-contractors to this type of contractual requirement is that the insurance should be maintained only as long as it remains available at a commercially reasonable rate of premium.

Professional indemnity insurance generally provides cover only in respect of a failure to exercise reasonable skill and care in the provision of design; it will rarely cover the more onerous liabilities sometimes imposed on designing sub-contractors such as an undertaking that the thing designed will be fit for its purpose.

However, sub-contractors such as liftmakers and structural steelwork contractors, who manufacture as well as fix the products they supply, may carry product liability insurance. This indemnifies them against liability for damage caused by defective products, including damage caused by products which are defective in design, and will cover damage caused because the product is unfit for its purpose.

In other ways, product liability cover is considerably more restricted than professional indemnity insurance. It will not, for example, cover design work carried out for a separate fee.

Of most significance to the employer will be the fact that it does not cover the cost of rectifying the defective products itself, only the consequential loss that the product causes. So if a defective lift falls to the bottom of the lift shaft, damage to the lift itself will not be covered, but any damage to the shaft and the surface onto which the lift falls would be insured. Even the cover for consequential loss may not extend to purely financial losses such as loss of profit or loss of rent – this will depend upon whether the sub-contractor has arranged an extension to his policy to cover such losses.

For these reasons, some employers are reluctant to accept product liability insurance as an alternative to professional indemnity insurance.

Bonds and guarantees

Since the 1980s, there has been a steady increase in requests for performance bonds and other guarantees of performance from contractors working on construction projects in the UK. Always a feature of overseas contracts, bonds are now prevalent on all major commercial developments, where they are required not only from the main contractor, but also from major sub-contractors.

Definition of a performance bond

The National Joint Consultative Committee for Building (NJCC) has defined a performance bond as:

> 'a three party contract between the employer, the contractor and the surety, guaranteeing performance by providing the employer with a stated maximum financial benefit in the event of non-performance'.

That definition relates to a bond provided by a main contractor. In the context of sub-contracts, it is of course the sub-contractor's performance which is guaranteed, and either the employer or the main contractor who will be the beneficiary, depending on how the bond is arranged.

The function of a performance bond is to indemnify the beneficiary (either the employer or the main contractor) against the consequences of the sub-contractor failing to perform his contractual obligations. If the sub-contractor is in default and the bond is called, the beneficiary may claim the sum bonded from the surety, provided the beneficiary has satisfied any terms which attach to the bond.

When the bond is called, the surety has a right to recover any sums he has paid under it from the sub-contractor under a separate agreement called a counter indemnity. The surety's only real risk is therefore that the sub-contractor will become insolvent. Not surprisingly, sureties tend to investigate very carefully the financial status of any sub-contractor whose performance they are asked to guarantee.

Types of performance bond

Although there is no universally accepted bond wording applicable to building work, bonds tend to follow a reasonably standard structure, and it is therefore tempting to assume that their effect is always the same. There are, however, some important variants in the wordings available, so any sub-contractor who is giving a bond will need to study its terms to assess what type of bond he is providing.

The main points which a sub-contractor being asked to provide a bond should bear in mind are as follows:

Is the bond on demand?

A true performance bond will provide that the surety is liable to pay the sum bonded only when the contractor is in default in the performance of his contractual obligations. If the bond says no more than that, it will usually be up to the beneficiary to prove that the sub-contractor is actually in default before the surety will be persuaded to pay, particularly if the surety is an insurance company rather than a bank.

Some bonds expressly provide that they may be called only on receipt of a specified form of proof of default, such as an architect's certificate or the award of a judge or arbitrator. Such clauses are generally to the sub-contractor's advantage, as they will discourage the beneficiary from calling the bond in the event of minor breaches of the sub-contract, and will lessen the extent to which the bond can be used as a 'sword of Damocles' hanging over the sub-contractor's head in any dispute involving allegations of non-performance.

An on-demand bond, on the other hand, is as onerous as it sounds. The employer has only to present the bond to the surety and it will be paid 'on demand' without the surety being required, or indeed permitted, to investigate the alleged breach by the sub-contractor.

It is accepted by most commentators that on-demand bonds are inappropriate to UK construction contracts, where the bond's major function is a protective one in the event of the sub-contractor's insolvency. On overseas contracts, on-demand bonds have a different role to play because the person performing the contract will often have no assets in the jurisdiction where he is working, so in the event of his default, the beneficiary has no other effective remedy than calling the bond.

Any sub-contractor who provides an on-demand bond should bear in mind that he is enabling the beneficiary to obtain the sum bonded simply by making a demand or asserting a breach of contract.

Notice requirements

The calling of a bond is a serious matter for any contractor or sub-contractor, and may well be a matter he would be asked to disclose to any insurance company from whom he might later seek a bond or other financial guarantee. Some bond wordings therefore provide for the beneficiary to give the sub-contractor written notice of his intention to call the bond so that the sub-contractor has an opportunity to put things right.

Sum bonded

The sum bonded is usually described somewhat inaccurately as a percentage of 'the sub-contract sum' – what is normally meant by this is the price the sub-contractor has tendered for the work. By convention, and under all of the JCT related forms of sub-contract, tender prices are calculated on a VAT-exclusive basis, so the sum bonded will also be calculated on the tender price net of VAT.

The percentage applied to the tender price to determine the sum bonded is usually 10 per cent. Although the NJCC recommends that 10 per cent is the maximum sum for which a bond should be required, demands for bonds of up to 15 or 20 per cent are becoming more common as the confidence of employers in the solvency of construction companies is eroded by a steady stream of insolvencies.

Termination dates

If the bond does not contain a termination date, the surety's liability under it will continue until the limitation period has expired on the contract in respect of which performance is being guaranteed. This will be six years if the contract is executed under hand, and twelve years if it is executed under seal or as a deed. This continuing exposure could lead to the surety demanding an additional fee or premium each year.

Most bonds do provide for the surety's liability to come to an end on a defined date, which may be the date of practical completion of either the sub-contract or the main contract works, or the end of the defects liability period, or the date of issue of the final certificate. From the sub-contractor's point of view the earlier the bond comes to an end the better, and so the date of practical completion of the sub-contract works would be preferable.

Types of surety

Bonds can be obtained either from banks or from insurance companies. The sub-contractor who is providing the bond will normally be left to decide from whom the bond should be obtained, and it is important that he exercises that choice with care as the source of the bond will often dictate both its terms and its cost.

Banks, for example, are not interested in becoming involved in disputes between the sub-contractor and the beneficiary of the bond. Their objective is to pay up if the bond is called, debit the sub-contractor's account with the sum they have paid and then drop out of the picture, leaving the sub-contractor to fight it out with the beneficiary if he thinks the bond has been called unjustifiably.

Banks therefore prefer on-demand bonds, and even if the bond is not itself on-demand, the sub-contractor is likely to find that the counter-indemnity required by the bank will permit them to pay out and to debit the sub-contractor's account on receipt of a demand, without making any further enquiries whatsoever.

Insurance companies, on the other hand, rarely give on-demand bonds, and usually investigate the reasons for any call on the bond before making a payment. They may even become involved in legal proceedings to determine whether the call on the bond is a proper one.

Banks charge a fee for the bonds they provide but, even more crucially for the sub-contractor, they will treat the sum bonded as part of the sub-contractor's overdraft facility. This means that the more bonds the sub-contractor gives, the more limited his credit facility, which may mean he is unable to capitalise his business.

Insurance companies charge a premium for the bonds they supply and, like the banks, require a counter-indemnity from the sub-contractor. However, they will not have the direct access to the sub-contractor's bank account which the bank enjoys, and therefore they may require the counter indemnity to be secured in some way, for example by a charge on the company's assets or by a personal guarantee from the directors.

Insurers also undertake quite detailed investigations into a sub-contractor's financial position before agreeing to give him a bond, because their main risk is the sub-contractor's insolvency. This means that, unless the sub-contractor already has a bonding facility in place, it can take a great deal of time to obtain a quotation for a bond, which can create problems in a short tender period.

Alternatives to bonds

Many contractors and sub-contractors strenuously resist requests for bonds, on the grounds that if proper financial checks are carried out at tender stage the risk of insolvency is minimal, and in the event of their default the beneficiary will have other remedies which will be quite sufficient provided the company remains solvent.

These arguments are sometimes successful, but there remains a substantial number of employers (and main contractors) who insist on the provision of some additional security for the sub-contractor's performance. If the sub-contractor is determined to avoid providing a bond, or if it simply not feasible for him to do so, there are other ways in which he can offer security for his performance.

If the sub-contractor is part of a group of companies, he may be able to offer a parent company guarantee. This will depend upon the willingness of the parent company to guarantee its subsidiary's performance, as some groups have a policy that each separate company should be entirely independent.

If such a guarantee is available, it offers real advantages to the sub-contractor compared with a bond because the parent company will not usually charge a fee, and the wording of the guarantee will usually give the sub-contractor an opportunity to remedy any non-performance before the guarantee is called.

Some parent companies will not give legally enforceable guarantees of the performance of their subsidiaries, but are prepared to provide a letter reassuring the other party that their subsidiary will be in a position to perform its contractual obligations. Such 'letters of comfort' do not impose any legal liability on the parent because there is no intention to create a legal relationship between the parent and the person seeking the guarantee, and they are therefore rarely acceptable if the employer or contractor is determined to obtain real security for the sub-contractor's performance.

Another alternative is to offer an increase in the amount of retention the contractor is entitled to deduct from the sub-contractor's interim payments. This is a simple way of giving the main contractor extra security, although it will obviously have cash flow implications for the sub-contractor, and may affect his ability properly to resource the contract. It may also not be the ideal solution where it is the employer, not the contractor, who is demanding additional security from the sub-contractor.

If the amount of retention is increased, the sub-contractor will need some comfort that the retention fund will be protected in the event of the insolvency of either the employer or contractor. The most effective way to obtain that protection is by payment into a separate trust account, or into a separate account in the joint names of the employer, contractor and sub-contractor.

11 Determining a Sub-Contractor's Employment

Even by JCT standards, the clauses relating to determination in the standard sub-contracts seem inordinately long and complex. But attempting to walk away from a sub-contract when you are not entitled to do so is an expensive exercise so they are, in fact, extremely important.

This chapter defines determination and the related concepts of forfeiture, frustration and repudiation, and outlines the contractual rules governing the right of both the main contractor and sub-contractor to determine the sub-contractor's employment. It also summarises the allocation between the main contractor and the employer of the risk of determination of the employment of nominated sub-contractors, named sub-contractors and works contractors.

Determination and forfeiture

Determination is the term used in the JCT standard forms and their related sub-contracts to describe the bringing to an end of the sub-contractor's right and obligation to carry out the sub-contract works under an express term of the contract.

It is important to note that it is generally the sub-contractor's employment which is determined, rather than the sub-contract itself, which ensures that the terms of the sub-contract dealing with the consequences of determination remain in force.

Clauses providing for determination are sometimes described by lawyers as 'forfeiture clauses', which is a general term used to denote the terms of a contract which permit the party paying for performance of the contract (in the context of sub-contracts, the main contractor) to determine the other party's employment or the contract itself, or to eject the sub-contractor from the site or otherwise to take the sub-contractor's work into their own hands.

Repudiation and frustration

In addition to the rights which arise under the express terms of the sub-contract, either party may also be entitled at common law to treat the sub-contract as being at an end if it has been repudiated or frustrated.

A party to a contract repudiates it where he makes it clear that he will not perform his contractual obligations, or where he acts in breach of a term which goes to the root of the contract (known as a fundamental breach). When a repudiation occurs, the 'innocent' party may either accept it, in which case the contract is at an end and he will be excused from further performance, or he may affirm the contract, in which case the repudiating party will remain under an obligation to perform. In both cases, the innocent party will also be able to recover damages for all of the loss he has suffered due to the other's breach.

Examples of circumstances where a sub-contract may be repudiated include a failure by the main contractor to pay the sub-contractor (although it must be a very serious failure which justifiably destroys the sub-contractor's confidence that he will ever be paid) or the sub-contractor wrongfully withdrawing from site.

In contrast to repudiation, frustration occurs where neither party is at fault, but the contract nevertheless becomes impossible to perform. This is a rare occurrence, and the cases where frustration has been held to have occurred often have curious facts – for example, when rooms were let to view the coronation procession of Edward VII which was postponed because of the King's illness, the contract of hire was held to be frustrated (*Chandler* v. *Webster* [1904] 1 KB 493).

A contract may also be frustrated when events occur which mean that the contractor is required to provide something different from that for which he originally contracted. In the words of Lord Radcliffe in *Davis Contractors* v. *Fareham Urban District Council* [1956] 2 All ER 145:

> 'frustration occurs whenever the law recognises that without default of either party a contractual obligation has become incapable of being performed because the circumstances in which performance is called for would render it a thing radically different from that which was undertaken by the contract'.

It will usually be extremely difficult to prove that a building contract has been frustrated. As explained in Chapter 6, a large number of variations will not be sufficient to constitute frustration, and in the *Davis* case cited above it was held that a scarcity of labour which resulted in completion taking twenty-two months rather than eight did not amount to a frustrating event.

Circumstances where frustration has been held to have occurred include a landslip which destroyed flats under construction where a permit to rebuild could not be obtained for a further three years (*Wong Lai Ying* v. *Chinachem Investment Co. Ltd* (1979) 13 BLR 81), and a three year delay in a contract to construct reservoirs due to the outbreak of the First World War (*Metropolitan Water Board* v. *Dick, Kerr & Co. Ltd* [1918] AC 119).

When a contract is frustrated, both parties are excused from further performance and, in principle, neither is liable to the other for any damage which results. However, the sub-contractor will have a statutory right under the Law Reform (Frustrated Contracts) Act 1943 to be paid for any 'valuable benefit' which the main contractor has received, which will be assessed by the court in accordance with what is just in the circumstances of the case. In practice, this will usually be calculated on the same basis as *quantum meruit.*

New model determination clauses for the JCT forms

The JCT has recently reviewed all of the determination clauses in its major standard forms with a view to ensuring that they are as clear as possible. In the course of that review, the Tribunal identified a number of inconsistencies between the clauses in the different forms, which it was felt were unjustified and confusing for users. A model set of determination clauses has therefore been produced which is gradually being incorporated into all of the standard forms and their related sub-contracts. When this process is complete, it will mean that once an employer, contractor or sub-contractor is familiar with the principles in one standard form, they can be confident that the determination clauses in the other contracts will be the same, except where differences in the forms mean that a deviation from the model clauses is essential.

At the time of writing in February 1994 the new model clauses have been incorporated into only JCT 80 (by Amendment 11, issued July 1992) and NSC/C (by Amendment 1, also issued July 1992) and the With Contractor's Design Form 1981 (by Amendment 7 issued January 1994). It is anticipated that they will be incorporated into all of the main and sub-contract forms covered in this book by the end of 1994, but readers should check up-to-date versions of the individual forms to see whether this has occurred.

There is one important substantive distinction between the new model determination clauses and the existing provisions on determination: the model clauses introduce an option for the employer to determine the main contractor's employment on his insolvency (unless it is full liquidation or bankruptcy) rather than providing for automatic determination.

Determination under the standard forms of sub-contract

Every standard form of sub-contract caters for:

- determination by the main contractor on the grounds of the sub-contractor's default or insolvency;
- determination by the sub-contractor on the grounds of the main contractor's default;
- determination of the main contractor's employment under the main contract.

It is worth remembering that the clauses dealing with determination will be construed strictly by the courts, so the contractual provisions must be followed in every particular in order to ensure that the determination cannot be challenged on the basis that the party attempting to determine is guilty of repudiation.

The rules in each of the sub-contracts are summarised later in this section, but in order to understand them fully, it is necessary to appreciate how widely the JCT forms define insolvency.

Insolvency

Insolvency is the generic term used to describe the situation where a person, firm or company is unable to pay its debts. There are a number of different legal procedures which creditors (or sometimes the debtor himself) can invoke when this occurs, or when financial difficulties occur which are short of full insolvency.

The 'insolvency events' which can give rise to a determination are very widely defined in the JCT forms and their related sub-contracts, as they cover virtually all of these procedures. Insolvency in the context of the JCT forms therefore encompasses:

- **bankruptcy**, which is the insolvency of individuals and partnerships;
- **compositions/arrangements with creditors**, where creditors agree to take a percentage of sums due to them in full and final settlement of their claims or agree to accept payment in instalments over an extended period — if the creditor is a company such arrangements must be approved by the court under the Companies Act 1985 or the Insolvency Act 1986 in order to constitute an insolvency event for the purposes of the JCT forms;
- **administration**, which involves the appointment of an administrator by the courts on the application of the company itself to perform a specific purpose, such as to rescue it from temporary cash flow problems. During an administration creditors' rights are temporarily suspended;
- **receivership and administrative receivership**, which generally occurs where a debenture holder (such as a bank) exercises its right to appoint an individual to manage the business to obtain repayment of the loan secured by the debenture;
- **winding up**, which is the procedure for putting a company into liquidation. It may be initiated by the shareholders themselves, in which case it is known as a members' voluntary winding up, or by the creditors either forcing the shareholders to wind up the company (a creditor's voluntary winding up) or applying the court to order a winding up (a compulsory winding up). All of these procedures will have the effect of triggering the

determination clauses except for a members' voluntary winding up which is purely for the purposes of reconstruction or amalgamation.

It is worth noting that in itself, insolvency would not bring the contract to an end at common law, although its likely effects – non-payment, closure of the site, etc. – would usually amount to a repudiation.

Nominated sub-contract NSC/C

As explained above, the determination provisions in the nominated sub-contract NSC/C were substantially redrafted by Amendment 1 issued in July 1992, and follow the model clauses which will eventually be adopted in all the JCT-related sub-contracts.

They begin with a general statement that the notices which play such an important part in the determination process must be in writing and given by actual delivery (i.e. physically handed to the recipient) or by registered post or recorded delivery. Notices given under the determination clauses will not be valid if served by fax: in view of the increasingly widespread use of fax as a method of communication this is a significant restriction.

Determination by the contractor

Grounds for determination
The main contractor may determine the nominated sub-contractor's employ-ment if, before practical completion, he commits what clause 7.1.1 describes as a 'specified default', provided the main contractor follows the procedure set out in the sub-contract.

The specified defaults are:

- wholly or substantially suspending work without reasonable cause;
- failing to proceed regularly and diligently with the sub-contract works without reasonable cause;
- failing to comply with an architect's instruction or a contractor's direc-tion to remove defective work, materials and goods provided that, as a result of that failure, the works are materially affected;
- any breach of clauses 3.13 or 3.14 regarding assignment and sub-letting.

It will usually be reasonably clear when the sub-contractor has committed a specified default, but in one area there remains considerable scope for argument. This is where it is alleged that the sub-contractor is not proceeding regularly and diligently with his work.

The equivalent clause in the main contract was recently considered by the courts in *West Faulkner Associates* v. *London Borough of Newham* (1992) 61 BLR 81. In that case, Judge John Newey QC held that the requirement to carry out works regularly and diligently meant that contractors should:

'go about their work in such a way as to achieve their contractual obligations. This requires them to plan their work, to lead and manage their
workforce, to provide sufficient and proper materials and to employ competent tradesmen, so that the works are fully carried out to an acceptable
standard and that all time, sequence and other provisions of the contract
are fulfilled.'

The quantity surveyor in the *West Faulkner* case had apparently advised the
architect that as long as the contractor had some men and materials on the
site, and was completing some of the work within the contract period, they
were not failing to proceed regularly and diligently. Although few commentators would agree with that advice, the judge in this case did define the
obligation on the contractor in broad and potentially onerous terms – would
the employment of one incompetent individual, for example, constitute a
specified default?

On balance, it is suggested that whether a sub-contractor is proceeding
regularly and diligently will always be a question of fact and degree, and whilst
minor management failures should not give rise to a right to determine, it will
be no excuse for the sub-contractor simply to ensure that men and materials
are present on site if they are not being properly and efficiently utilised.

The procedure which the main contractor must follow if he wishes to
determine a nominated sub-contractor's employment on the grounds of a
specified default is set out in clause 7.1. It is far from straightforward.

Firstly, the main contractor must inform the architect of the default, and
pass to him any written observations of the sub-contractor on that default,
because it is the architect who decides whether the process should be set in
motion. If the architect decides that determination is appropriate, he instructs
the contractor under clause 35.24.6.1 of the main contract to issue a notice to
the sub-contractor specifying the default. In that instruction he should also
tell the main contractor whether he must seek a further instruction before
actually determining the sub-contractor's employment if the default is not
remedied or if it is repeated.

If the sub-contractor continues the default for 14 days after receipt of the
initial notice, the main contractor may determine his employment by issuing a
further notice (having first sought an instruction from the architect if the
architect has so stipulated in his initial instruction). This second notice must
be issued within 10 days of the end of the 14 day period for which the default
continued. This ensures that the main contractor must act reasonably quickly,
and the sub-contractor will not have the spectre of determination hanging
over him indefinitely.

If the sub-contractor rectifies his default after an initial notice has been
issued, or for some reason the main contractor does not issue the second
notice which actually determines the sub-contractor's employment within
the specified period, the sub-contractor remains at risk if he repeats a

specified default. In those circumstances, the contractor may determine the sub-contractor's employment either immediately upon or within a reasonable time of the repetition, having first sought an instruction from the architect if required to do so under the architect's original instruction.

This complex mechanism ensures that the architect is involved in the initial stages of any determination procedure, but allows him to leave the subsequent decisions to the main contractor if he wishes to do so.

This section of the determination provisions ends with a general caveat that the notice which actually determines the sub-contractor's employment must not be given unreasonably or vexatiously. These words were considered by the Court of Appeal in *John Jarvis* v. *Rockdale Housing Association* (1987) 36 BLR 48, when it was held that a notice would not be unreasonable unless it was issued in circumstances in which no reasonable contractor would have done so. Vexatious was said to imply 'an ulterior motive to oppress, harass or annoy'.

In the event of a serious default by the sub-contractor which does not fall within these express provisions, it should be borne in mind that the contractor's common law rights are expressly preserved by clause 7.6, therefore if the sub-contractor has repudiated the sub-contract it can be brought to an end without the need to rely on the determination provisions. This right can also be useful where the procedural requirements of the determination clauses are inappropriate or have been ignored.

The consequences of the insolvency of the nominated sub-contractor are dealt with in clause 7.2.

If the sub-contractor becomes bankrupt, has a provisional liquidator appointed, passes a resolution for a voluntary winding up (except for the purposes of reconstruction or amalgamation) or has a winding up order made against him by a court, his employment will be automatically determined, without the need for the architect or main contractor to take any action whatsoever.

If, however, the sub-contractor suffers one of the other insolvency events contemplated by NSC/C, such as the appointment of an administrator or administrative receiver, the contractor has the option (which he may only exercise with the written consent of the architect) to determine the sub-contractor's employment. If he wishes to do so, he simply issues a notice to the sub-contractor, and determination will occur on the date of its receipt.

A distinction has been drawn between liquidation and bankruptcy and other types of insolvency procedure in recognition of the fact that procedures such as voluntary arrangements and administration are designed to rescue a company from financial difficulties. This can be prejudiced if all of the company's contracts come to an abrupt end when, for example, an administrator is appointed, because once determination has occurred, it is more difficult to arrange for the sub-contractor to complete the sub-contract works, or for novation to a new sub-contractor.

The final ground for determination of the sub-contractor's employment, if the employer (**not** the architect) so requires, is corruption by the sub-contractor. NSC/C does not stipulate any particular procedure for such a determination, so it is presumably effected by a simple notice.

Consequences of determination by the main contractor
Where the main contractor determines the nominated sub-contractor's employment on any of the grounds set out above, clause 7.5 of NSC/C governs the rights and obligations of the parties.

The provisions of that clause may be briefly summarised as follows:

- any replacement nominated sub-contractor appointed by the architect may carry out and complete the sub-contract works and may use the departed nominated sub-contractor's temporary buildings, plant and equipment, and any goods and materials on site (provided that if they are not owned by the contractor or departed sub-contractor the owner's consent has been obtained);
- provided the sub-contractor is not insolvent, and if so required by the employer within 14 days of the determination, the sub-contractor must assign sub-sub-contracts and contracts for the supply of materials to the contractor to the extent they are assignable (they will often contain an express prohibition on assignment without consent);
- provided the sub-contractor is not insolvent, the contractor must pay sub-sub-contractors and materials suppliers direct if directed to do so by the architect;
- when instructed to do so, the sub-contractor must remove his temporary buildings, plant, equipment and goods and materials from the site – if he does not respond to that instruction the contractor may remove and sell them and hold the proceeds to the credit of the sub-contractor (but less his own costs and without responsibility for loss or damage).

Perhaps the most important aspect of clause 7.5.3 is that, once a determination has occurred, the contractor is not obliged to make any further payments to the nominated sub-contractor until after the sub-contract works have been completed and the defects liability period has expired. The new model clauses introduce an exception to this rule where the main contractor has unreasonably failed to pay amounts which 'accrued' – i.e. fell due – 31 days or more before any insolvency event. This aspect of the clause is curiously worded, as it is not entirely clear whether the right to pursue old debts only survives where insolvency was the reason for the determination, although it seems likely that this was the intention.

Once the defects liability period has expired, the sub-contractor may apply to the main contractor for payment. The architect will ascertain the expenses and loss and expense suffered by the employer as a result of the determination, which are then deducted from any sums due to the sub-contractor in respect of work or materials which had not been paid for at the date of the determination. The contractor may also deduct his own loss and expense and a cash discount of two and a half per cent. If there are no further sums due to the sub-contractor once the employer has deducted his loss and expense, the contractor will be entitled to recover his own loss and expense as a debt, although that will be of little use if the sub-contractor is insolvent.

Determination by the sub-contractor

A nominated sub-contractor has fairly restricted rights to determine his own employment, but as his common law rights are preserved by clause 7.9 they are not an exhaustive remedy. He also has the very valuable protection of being able to suspend work for non-payment, which is discussed in Chapter 7.

Grounds for determination
The sub-contractor may initiate the determination process where without reasonable cause the main contractor has either:

- wholly or substantially suspended the carrying out of the main contract works; *or*
- failed to proceed with the main contract works so that the reasonable progress of the sub-contract works is seriously affected.

The procedure which the sub-contractor must follow is very similar to that laid down in relation to determination by the main contractor. The sub-contractor issues a notice that a 'specified default' has occurred, and if it continues for 14 days the sub-contractor then has a further 10 days to issue a second notice to the contractor, receipt of which determines the sub-contractor's employment.

If the contractor ends the specified default within 14 days of the issue of the first notice, or if the sub-contractor does not issue the second notice, he may nevertheless determine his employment by a further notice if that particular default is repeated.

The sub-contractor is under a similar obligation to the main contractor not to issue a notice of determination unreasonably or vexatiously; for an explanation of that phrase, see above.

Consequences of determination by the sub-contractor
The rights and obligations of the parties where the sub-contractor has determined his own employment are set out in clause 7.8.

Firstly, he must 'with reasonable dispatch' remove from the site his temporary buildings and plant and unfixed goods and materials and ensure that his sub-sub-contractors do the same. This is a fertile area for disputes as to whether property in goods and materials on site has passed to the employer, in which case they cannot be removed.

The sub-contractor may need to be reminded of clause 4.15.4.2 of NSC/C, which provides that where the value of unfixed goods and materials has been included in a certificate and that certificate has been paid or otherwise discharged by the employer, property in those goods and materials passes to the employer, and the sub-contractor expressly undertakes not to challenge the employer's title.

Another relevant proviso in these circumstances is clause 7.8.2.4, which states that one of the sums due to the sub-contractor on determination is the cost of goods and materials for the sub-contract works for which the sub-contractor has paid or is legally bound to pay. If the contractor pays in full for those goods and materials, they become his property and therefore should not be removed from site.

Given that relationships are likely to have deteriorated quite seriously by the time the sub-contractor determines his own employment, difficulties may arise where the sub-contractor has paid his suppliers (or is bound to do so) but has no real confidence that the contractor will honour his obligation to pay under clause 7.8.2.4. In those circumstances, it is suggested that the sub-contractor would be within his rights to remove the relevant goods and materials from site, and return them only when he has been paid by the main contractor.

Having removed his property from the site, the sub-contractor is required to prepare an account and submit it to the contractor setting out:

- any agreed amount of loss and expense due to disruption caused by the contractor's default;
- the reasonable cost of removing his property from site;
- loss or damage he has suffered due to the determination;
- the cost of goods and materials properly ordered for which he has paid or is legally bound to pay (note they do not have to have been delivered to the site).

This account should be paid by the contractor within 28 days of its submission, without deduction of retention or cash discount. The contractor may, however, deduct the amount of any relevant previous payments and any agreed loss and expense suffered by the main contractor due to the sub-contractor's default.

In addition, the sub-contractor will be entitled to be paid any sums certified in his favour by the architect in the normal way.

Effect of determination of main contractor's employment

When the main contractor's employment is determined under the main contract, for whatever reason, the sub-contractor's employment will also automatically determine.

The sub-contractor must then remove his temporary buildings, plant, goods and materials from the site and ensure that his sub-sub-contractors do the same.

His entitlement to be paid, however, will vary depending upon whether the main contractor's employment was determined on the grounds of his own default or insolvency under clause 27 of JCT 80, or because of the employer's default or insolvency under clause 28 or due to one of the neutral events under clause 28A.

Where the determination of the main contractor's employment is due to his insolvency or default, the sub-contractor prepares his own account of sums due to him, which should comprise:

- the total value of work properly executed up to the date of the determination;
- any loss and expense due to disruption of the sub-contractor's regular progress due to the relevant matters listed in clause 4.38 of NSC/C which has been ascertained by the architect;
- any agreed amount of loss and expense due to disruption caused by the main contractor's default;
- the cost to the sub-contractor of removing his property from the site;
- direct loss or damage suffered due to the determination (eg loss of profit on the balance of the sub-contract work);
- the cost of materials properly ordered for the sub-contract works for which the sub-contractor has either paid or is legally bound to pay.

The sub-contractor is entitled to be paid these amounts within 28 days of the submission of his account, less any previous payments but without deduction of retention or cash discount.

This right will, of course, only be valuable if the main contractor has the money to pay the sub-contractor's account. If he is insolvent, it will be of very little use, although the sub-contractor may have a right to direct payment in respect of any sums certified by the architect, provided the main contractor's insolvency does not amount to bankruptcy or liquidation. The rights of a nominated sub-contractor to direct payment are discussed in Chapter 7 on Payment.

Where the contractor has determined his own employment under the main contract due to the employer's default or insolvency, the nominated sub-contractor is entitled to be paid the following amounts:

- any sums certified in favour of the nominated sub-contractor by the architect, provided the certificate was issued before the determination of the main contractor's employment;
- the part of any payments made to the main contractor by the employer following the determination which is fairly and reasonably attributable to the sub-contract works;

- any nominated sub-contract retention released to the main contractor after the determination.

The contractor must make the latter two payments within seven days of receiving the money from the employer, and may deduct cash discount if he pays within this period.

These rules mean that the nominated sub-contractor shares with the main contractor the risk of the employer becoming insolvent whereas, in theory at least, domestic sub-contractors employed under the standard forms retain their rights to be paid in full for the work they have carried out, notwithstanding the insolvency of the employer.

Where the main contractor's employment has been determined due to the occurrence of one of the neutral events listed in clause 28A, the employer is obliged to pay the contractor all of his costs, including any loss or damage he has suffered due to the determination, and the employer must inform the contractor in writing of how much of that payment is due to the nominated sub-contractor.

The main contractor must also pass on any nominated sub-contract retention released by the employer, and must pay in full any sums certified in the nominated sub-contractor's favour before the date of the determination. The latter payment must be made in accordance with the normal payment rules, i.e. within 17 days of the issue of the certificate, whilst the other two sums must be paid within 7 days of receipt of the money by the main contractor, who may deduct a cash discount provided he pays within the stipulated period.

Employer's option to determine – effect on nominated sub-contractors

Before the introduction of the employer's option to determine on the main contractor's insolvency, if the main contractor became insolvent his employment under the main contract automatically determined and therefore the employment of all sub-contractors was also automatically brought to an end. This would obviously be inappropriate in the light of Amendment 11 to JCT 80, as automatic determination of all the sub-contracts would probably be fatal to any novation or continuation of the project. Amendment 1 to NCS/C therefore introduces new provisions which operate when, due to an insolvency event, the employer is entitled to determine the main contractor's employment but has not yet done so.

In those circumstances, the contractor must inform the nominated sub-contractor that an insolvency event has occurred, for example that an administrative receiver has been appointed, but the sub-contractor has no right to determine at that stage. However, under clause 7.12.2 his obligation to carry out and complete the work is suspended until one of the following events occurs:

- the sub-contractor reaches an agreement with the main contractor for the continuation of the sub-contract works; *or*
- the employer exercises his option to determine the main contractor's employment; *or*
- a period of three weeks (or such additional period as may be agreed) has expired.

The sub-contractor is entitled to an extension of time and loss and expense for this period of suspension, but as it is deemed to have been caused by a default of the main contractor his claim will be against the main contractor, which will, of course, be of limited value as the main contractor is insolvent.

These arrangements give the contractor's insolvency practitioner and the employer at least three weeks to negotiate arrangements for the completion of the project, and the sub-contractors will generally play a vital role in those negotiations. If they are not involved, they have the right to walk away from the job once the three week period of suspension has expired. This is bad for everyone, as it will hinder the employer's plans for completion and will lessen the sub-contractor's chances of payment.

Although nominated sub-contractors are, strictly speaking, entitled to direct payment where the main contractor's insolvency falls short of full liquidation or bankruptcy, many employers are reluctant to make direct payments in those circumstances in case they are challenged by the receiver or administrator. This means that, in practice, the sub-contractor will usually be in a better position if he is involved in the completion of his work, so the arrangements described above will generally operate in his favour.

Domestic sub-contract DOM/1

DOM/1 has not yet been amended to reflect the inclusion of the new model determination clauses in JCT 80, but its clauses on determination are broadly similar to those in NSC/C save for the fact that where the employer has an option to determine under the main contract, there are no express terms to deal with the position of the sub-contractor where the main contractor has become insolvent but the employer has not yet exercised his option to determine.

Determination by the main contractor

The contractor is entitled to determine the domestic sub-contractor's employment if:

- he wholly suspends the carrying out of the sub-contract works without reasonable cause; *or*
- he fails to proceed with the works in accordance with the details in the Appendix and reasonably in accordance with the progress of the main

contract works (note that the phrase 'regularly and diligently' is not used in DOM/1); *or*
- he refuses to remove defective work or materials or refuses to remedy defects and as a result the works are materially affected; *or*
- he fails to comply with clause 26, which deals with assignment and sub-letting.

The procedure for determination is less complicated than that in NSC/C. The main contractor simply issues a notice to the sub-contractor by registered post or recorded delivery specifying the default, and if it is not remedied within 10 days of receipt of the notice, or is repeated at any time, the contractor has a further 10 days in which to issue a further notice actually determining the sub-contractor's employment. The notice of determination itself must not be given unreasonably or vexatiously: for guidance on the meaning of this phrase, see under the commentary on NSC/C.

If the domestic sub-contractor becomes insolvent (which, as under NSC/C, includes voluntary arrangements, receivership and administration) the contractor may simply determine the sub-contractor's employment by written notice.

Similarly, if the sub-contractor is guilty of corruption as defined in clause 29.3 of DOM/1, the main contractor may determine his employment by a single notice.

Note that there is no need for any architect's instruction or permission where the main contractor wishes to determine a domestic sub-contractor's employment: it is entirely a matter between the two parties to the sub-contract.

Consequences of determination by the contractor
Where the contractor has determined the domestic sub-contractor's employment, clause 29.4 of DOM/1 provides that:

- the contractor or other sub-contractors may use all the sub-contractor's temporary buildings, plant and tools, goods and materials on site;
- provided the determination was not due to the sub-contractor becoming bankrupt or being wound up, the sub-contractor must, if required to do so by the main contractor, assign to him all sub-sub-contracts and supply contracts but on terms that the sub-sub-contractors and suppliers have a right of reasonable objection to any further assignment by the contractor (although legally it may be impossible for the sub-contractor to comply with this requirement if his sub-sub-contractors and suppliers are entitled by their contracts to withhold their consent to an assignment);
- the main contractor may also pay sub-sub-contractors and suppliers direct provided the determination was not due to bankruptcy or wind-

ing up, but only to the extent that payments have not already been made to the original sub-contractor;

• when required to do so by the main contractor, the sub-contractor must remove his temporary buildings, plant and equipment from the site. If he does not do so, the contractor may remove and sell them without responsibility for loss or damage, and holds the proceeds of sale to the credit of the sub-contractor, less his own expenses.

As under NSC/C, the sub-contractor is not entitled to receive any payment from the contractor until completion of the sub-contract works (not the main contract works). This is obviously an important distinction where the sub-contractor whose employment has been determined is an early trade such as piling or steelwork. There has been some debate as to whether 'completion' in this context means practical completion or completion of making good defects. In *Emson Eastern Ltd* v. *EME Developments Ltd* (1991) 55 BLR 114 it was held that it was not realistic to separate the concepts of completion and practical completion, and therefore payment should be made after practical completion of the sub-contract works.

In practice there will usually be little money due to the sub-contractor after a determination due to his default or insolvency, as the main contractor is entitled to deduct both cash discount and any loss or damage he suffered due to the determination (which will include the additional costs he incurred in selecting and employing someone else to complete the project) from the value of work and materials supplied by the sub-contractor for which he had not been paid at the date of the determination.

Determination by the domestic sub-contractor

Grounds for determination
A domestic sub-contractor has slightly wider rights to determine his own employment than a nominated sub-contractor, as in addition to the grounds of suspension of work before practical completion and failure to proceed, the domestic sub-contractor may also determine if the main contractor fails to pay in accordance with the sub-contract. This right has presumably been omitted from the nominated sub-contract because of the sub-contractor's rights to direct payment under NSC/C.

In any event both sub-contracts permit suspension for non-payment, which is usually a less drastic remedy than determination, and can be very effective. However, the existence of the right to suspend may limit the domestic sub-contractor's right to determine, as he may only determine where the sub-contract does not provide any other adequate remedy for the main contractor's default. If applied strictly this could mean that the sub-contractor can only determine for non-payment in special circumstances, for example where he has an expensive establishment on site which he could remove on determination but which would have to stay on site where he merely suspended.

The procedure for determination by the sub-contractor is the same as for determination by the main contractor: the sub-contractor issues a notice of default to the main contractor by registered post or recorded delivery, and may determine after 10 days if the default is not remedied or is repeated at any time.

If the sub-contractor has elected to suspend work for non-payment under clause 26.1, he may not issue a notice of determination until 10 days after the suspension has begun. This means that if the sub-contractor feels the main contractor's default is sufficiently serious to warrant determination rather than suspension, it is important that his notice to the main contractor makes his intentions absolutely clear.

If the main contractor becomes insolvent, that in itself does not entitle the sub-contractor to determine his own employment. However, if the main contractor's employment is determined under the main contract (which will be at the employer's option following the introduction of Amendment 10 into JCT 80) that will automatically determine the sub-contractor's employment. Alternatively, insolvency is likely to lead to suspension of the works, thus giving the sub-contractor the right to determine, although he will still have to follow the procedure of giving 10 days notice of the default described earlier.

Consequences of determination by the sub-contractor

Where the sub-contractor determines his own employment, he must remove his temporary buildings, plant and equipment from the site, together with those goods and materials which he owns. The considerations which apply here are identical to those discussed above in relation to nominated sub-contracts.

On the determination of his employment, the domestic sub-contractor is entitled to be paid:

- the total value of work executed up to the date of the determination, with any uncompleted work being valued as if it were a variation;
- any loss and expense incurred by the sub-contractor due to the disruption of his regular progress;
- the cost of materials and goods for which the sub-contractor has paid or is legally bound to pay;
- the reasonable cost of removing his temporary buildings, etc.

Note that there is no express right to recover loss and expense caused by the determination itself, which would include sums such as loss of profit. However, the sub-contractor's rights under the express terms of the contract are stated to be without prejudice to his common law rights, and so he should be able to recover such sums.

Effect of determination under the main contract

If the main contractor's employment is determined for any reason whatsoever, the sub-contractor's employment automatically determines under clause 31 of DOM/1. In that event, the sub-contractor's rights are the same as where he determines his own employment. However, if the reason for the determination under the main contract does not constitute a breach of the sub-contract, the sub-contractor will have no common law right to loss and expense due to the determination.

Domestic sub-contract DOM/2

The rules relating to determination in DOM/2 are precisely the same as those in DOM/1.

Named sub-contract NAM/SC

The determination clauses in NAM/SC are almost identical to those in DOM/1, save for the following:

- where the sub-contractor's employment is determined by the main contractor, there are no express provisions requiring the assignment of sub-sub-contracts or contracts of supply and no provisions for direct payment to suppliers and sub-sub-contractors;
- where the sub-contractor determines his own employment or where it is determined due to the determination of the main contract other than for the employer's default, the sub-contractor has an express right to recover any loss and expense he suffers due to the determination.

Domestic sub-contract IN/SC

The provisions governing determination in IN/SC are identical to those in NAM/SC.

Works Contract/2

Section 7 of Works Contract/2 sets out the rules regarding the determination of the works contractor's employment, which are very similar to those of DOM/1. There are, however, certain material differences, which are as follows:

- if either party wishes to determine on the grounds of default by the other, 14 days notice must be given rather than the 10 days required by DOM/1;

- the architect must be kept informed if the management contractor decides to determine the works contractor's employment on the grounds of default, although his consent is not required;
- there is an express reference to reinstatement of the works contractor's employment in the event of an automatic determination due to the works contractor's insolvency, provided the employer, the management contractor, the works contractor and his insolvency practitioner all agree;
- the management contractor may determine the works contractor's employment on the grounds of corruption only where the employer so requires;
- the amount due to the works contractor following a determination due to his default or insolvency is ascertained by the architect (or the quantity surveyor where so instructed by the architect) and included in an interim certificate – the management contractor may deduct cash discount and his own loss and expense from the certified sum;
- if the works contractor has suspended work on the grounds of non-payment, he may not issue a notice of determination for non-payment until 14 days after the suspension began;
- the works contractor has an express right to recover loss and expense arising out of the determination where he has determined his own employment.

Allocation of risk of determination of the employment of employer-selected sub-contractors

A major bone of contention between employers and main contractors is the extent to which the employer should bear the risk of the insolvency or default of the sub-contractors he has selected. Many employers start from the premise that the main contractor should be entirely responsible if any sub-contractor defaults, but this approach has been tempered in the JCT standard forms in relation to nominated sub-contractors under JCT 80, named sub-contractors under the Intermediate Form and works contractors under the JCT Management Contract.

These forms vary in the way in which they allocate the risk of default, but all of them provide for it to be shared between the contractor and the employer. The precise division of responsibility is described below.

Nominated sub-contractors under JCT 80

In principle, clause 35 of JCT 80 places upon the employer the risk of determination of the nominated sub-contractor's employment due to his insolvency or default.

It is likely that this would be the position at common law: in *Bickerton (T.A.) & Son* v. *North West Metropolitan Regional Hospital Board* [1970] 1 All ER 1039 the House of Lords held that where a sub-contractor nominated under JCT 63 had repudiated his sub-contract the employer was bound to nominate a new sub-contractor and to bear all of the increased costs incurred by the main contractor as a result of such re-nomination, including damages in respect of any delay caused by waiting for an instruction to renominate, even though there were no express terms in the main contract to that effect.

Express provisions are now included in clause 35 of JCT 80 requiring the architect to renominate whenever a nominated sub-contractor's employment is determined, although this may be postponed if the determination was due to the sub-contractor going into receivership or administration, provided there are reasonable grounds to suppose that the original sub-contractor is prepared to complete his work on terms which will not prejudice the interests of the parties (clause 35.24.7).

The architect must issue the re-nomination instruction within a reasonable time, the length of which will obviously depend on the effect of the nominated sub-contract work on the progress of the whole project. However, even when a sub-contractor's employment is determined at a critical stage, it is suggested that any court called upon to assess what constitutes a reasonable time would take into account the need for the employer to seek competitive tenders for the sub-contract to complete the works, and therefore a delay of several weeks may occur. Unless the main contractor could prove that this was unreasonable in all the circumstances, it is submitted that he would not be entitled to an extension of time for that delay, and thus would be exposed to a claim for liquidated damages by the employer.

If, however, the replacement nominated sub-contractor requires a longer period that the original sub-contractor to complete the nominated sub-contract work, this should entitle the contractor to an extension of time under clause 25.4.5.1 on the basis that it is a delay caused by an instruction issued under clause 35.

The work for which a new sub-contractor is nominated must include the rectification of any defects in the work executed by the original sub-contractor; this was confirmed by the courts in the case of *Fairclough Building Ltd* v. *Rhuddlan Borough Council* (1985) 30 BLR 26, and is now reflected in the express wording of clause 35. However, the employer gets a credit from the main contractor in respect of the amounts paid to him in respect of the original nominated sub-contractor's work which has now proved to be defective.

It is important to note that the amount of that credit is based on the original nominated sub-contractor's price rather than on the cost of rectifying the defective work. The latter may well be greater due to either the effects of inflation, underpricing of the original work or the complexity of rectification compared to the cost of executing the work properly in the first place.

A further exception to the general rule that the employer bears the costs of renomination is where the original sub-contractor has determined his own employment on the grounds of the main contractor's default. In those circumstances, the architect must still renominate within a reasonable time, but any additional amounts payable by the employer as a result of such renomination may be deducted by the employer from payments due to the contractor or may be recovered by the employer as a debt.

Named sub-contractors under IFC 84

The provisions which apply on the determination of a named sub-contractor's employment under IFC 84 are enormously complicated, and reflect the conflict between the employers' desire to bear as little of the risk as possible and the main contractors' determination that the power of selection must bring with it certain responsibilities.

The rules in clause 3.3.3 of IFC 84 are quite different to those in clause 35 of JCT 80, partly because named sub-contract work is not the subject of a prime cost sum but is priced by the main contractor in the ordinary way. This means that when a named sub-contractor's employment is determined, the architect is not obliged to rename but has a range of options. When the determination of a named sub-contractor's employment takes place, the main contractor must notify the architect, who is then obliged to issue instructions either:

- to name another sub-contractor to complete the work; *or*
- to require the main contractor to make his own arrangements for the completion of the work (which would include letting it to a domestic sub-contractor); *or*
- to omit the outstanding balance of the work (in which case the employer could employ the sub-contractor directly to complete the rest of the work).

The allocation of the cost and time consequences of complying with such an instruction will vary according to whether the sub-contractor was named in the main contract documents or in an instruction to expend a provisional sum.

If the sub-contractor was named in the main contract documents and the architect names a replacement sub-contractor, the contractor is entitled to an extension of time for any delay caused by the renaming, but not to loss and expense. The contractor will also recover any additional sums charged by the new named sub-contractor over and above those which would have been payable to the original named sub-contractor, save that the contractor must give a credit for any sums paid to the original sub-contractor in respect of work which was defective.

If the architect omits the balance of the named sub-contract work or instructs the contractor to complete it himself, that instruction will be treated as a variation and so the contractor will be paid all of the costs he incurs as a result of complying with it, and will get an extension of time and loss and expense if appropriate.

If the original sub-contractor was named in an instruction to expend a provisional sum, and therefore the main contractor was not aware of his identity when he tendered for the project, he is almost completely protected, as the architect's instruction regarding the balance of the work will be treated as a further instruction under the provisional sum and will therefore entitle the main contractor to recover his costs, an extension of time and loss and expense.

Although the main contractor may determine the named sub-contractor's employment without the architect's consent, he may only determine in accordance with the provisions of clauses 27.1 and 27.2 of NAM/SC, which permit determination on the grounds of certain specified defaults or the sub-contractor's insolvency. The main contractor is prohibited by clause 3.3.3 of the main contract from exercising any common law right he may have either to determine the sub-contract or to treat it as repudiated by the sub-contractor. If he breaches this prohibition, the architect must issue instructions indicating what is to happen to the balance of the named sub-contract work, but the contractor will not be entitled to an extension of time or to loss and expense as a result of complying with that instruction, and there will be no adjustment to the contract sum unless it is a reduction.

Where the named sub-contractor's employment has been determined on the grounds of his default or insolvency, the main contractor must pursue him to recover any costs arising out of the determination. The main contractor need not, however, go as far as commencing arbitration or litigation unless the employer agrees to indemnify him against his legal costs.

IFC 84 states that these sums include any liquidated damages which the employer would have been entitled to deduct had the main contractor not been entitled to an extension of time under the main contract. Clause 27.3.3 states that the sub-contractor must not try to argue that the main contractor is not entitled to recover such sums because he is relieved of liability under the main contract and has therefore suffered no loss. This is a similar arrangement to the relief provisions in the JCT Management Contract, and although it is understood that these clauses were drafted on counsel's advice, they have yet to be tested in the courts and it is not therefore certain whether they will be effective.

Works contractors under the JCT Management Contract

The JCT Management Contract is designed to be a low risk undertaking for the management contractor, and therefore the employer bears virtually all of the risk in the event of the determination of a works contractor's employment due to his default or insolvency.

Although clause 3.21.1 of the Management Contract provides that the management contractor must operate the terms of the works contract to recover sums due from the defaulting works contractor and must secure completion of the project by employing replacement works contractors as necessary, clause 3.21.2 entitles the management contractor to payment of all amounts which he properly incurs in fulfilling those obligations.

This should include any legal costs incurred by the management contractor in pursuing the defaulting works contractor, although the opening words of clause 3.21.1 require the management contractor to take the necessary steps 'in consultation with the architect and the employer' and in practice it would be foolish to commence legal proceedings without their knowledge and consent.

The 'relief provisions' in this clause also state that the employer may only recover liquidated damages for any delay caused by a works contractor's default (including a default which results in determination) to the extent that the management contractor is able to recover those sums from the defaulting works contractor.

Although the management contractor must meet claims made by other works contractors due to the determination of a works contractor's employment, if he cannot recover those sums from the defaulting works contractor it is the employer rather than the management contractor who bears the shortfall.

12 Methods of Dispute Resolution

Building is a long and complex process which involves a great deal of money and a multiplicity of parties who often have directly conflicting interests, so it is hardly surprising that it gives rise to some fierce disputes. Although it would be wrong to assume that the construction process is inevitably fraught with conflict, it is an industry which has generated more than its fair share of legal proceedings in recent years.

This chapter is therefore concerned with the different methods of dispute resolution available to the parties to a building contract: arbitration, litigation (i.e. court proceedings) and adjudication. In addition, various mechanisms for assisting the parties to reach their own settlement through the intervention of a neutral adviser are becoming increasingly popular in the UK. The generic name given to these methods is alternative dispute resolution (ADR), and the most popular are described at the end of the chapter.

Arbitration

All of the standard forms of sub-contract for use with the main contracts published by the JCT contemplate that any disputes or differences arising out of the sub-contract or in connection with it will be referred to arbitration. In practice, the parties do have a measure of control over the method used to determine disputes which may arise – some of the ways in which disputes can be brought before the courts, for example, are discussed under the heading 'Litigation'.

However, arbitration remains one of the most common methods used in the building industry to resolve disputes. This section briefly summarises the JCT rules relating to arbitration and addresses some of the matters which are particularly relevant to sub-contracts, such as joinder. A detailed explanation of arbitration procedure and the extensive legislation which controls it is beyond the scope of this book: there are many standard textbooks dealing with the subject, notably Mustill and Boyd on *Commercial Arbitration* and *Construction Arbitrations – A Practical Guide* by Vincent Powell-Smith and John Sims.

Arbitration clauses in JCT related sub-contracts

The articles of agreement of the JCT related sub-contracts all contain a brief agreement to submit all disputes and differences to arbitration, and the more detailed provisions are set out at the end of the conditions. This structure has been adopted because the Arbitration Act 1950 defines an arbitration agreement as 'a written agreement to submit present or future differences to arbitration, whether an arbitrator is named therein or not.' As all of the standard sub-contracts now incorporate the conditions by reference, it is questionable whether an arbitration agreement which appeared only in the conditions would constitute the 'written agreement' required by the Act. Incorporating a brief arbitration agreement in the articles of agreement, however, ensures that the Arbitration Acts will apply.

The arrangements for commencing an arbitration are reasonably straightforward: either side may give notice to the other that they wish to refer the dispute to arbitration. This 'notice of arbitration' is an important document: it stops time running for the purposes of limitation, and it defines the dispute which the arbitrator is empowered to determine. It is therefore vital that the notice describes the dispute as broadly as possible, to ensure that the respondent cannot subsequently argue that any matter raised in the points of claim is outside the scope of the notice. For this reason, it is usual for lawyers or other professional advisers to draft the notice of arbitration, although this is not a requirement of either the arbitration clause or the general law.

If the parties cannot agree on the identity of an arbitrator within 14 days of the date of the notice, either of them may request the appointer named in the appendix to the sub-contract to select an appropriate individual. The appendix provides for the appointer to be the President or a Vice-President of the RIBA, RICS or the Chartered Institute of Arbitrators (CIArb). If no appointer has been selected in the appendix, it provides for the President or a Vice-President of the RICS to make the appointment.

Agreement on the identity of the arbitrator is usually preferable to an appointment by an external body: even if relationships have deteriorated to the extent that it is difficult for the parties to agree on anything, losing the right to select an appropriate arbitrator means that an important benefit of arbitration is lost.

The sub-contracts for use with the Intermediate Form 1984 – NAM/SC and IN/SC – permit arbitration to be opened at any time on any issue. They are, however, unique in this respect. The other JCT related sub-contracts do not permit a reference to arbitration to be opened until after actual or alleged practical completion, termination or abandonment of the main contract works, except in relation to certain specified matters. This embargo has been imposed to ensure that the attention of the parties is not diverted from completing the works, although any really serious dispute is likely to have an adverse effect on the rest of the project regardless of whether formal arbitration proceedings are actually underway.

The matters on which arbitration under the sub-contracts can be opened before practical completion encompass some, but not all, of the issues on which at least one of the parties will want an immediate decision. They are:

- whether a payment has been improperly withheld or is not in accordance with the sub-contract;
- whether the sub-contract works are practically complete;
- matters arising out of the award of an adjudicator in respect of a dispute on the set-off provisions;
- whether the sub-contractor has made a reasonable objection to a variation in the restrictions or obligations under which the sub-contract works are to be carried out;
- matters relating to the powers of the main contractor to issue instructions following the execution of defective work by the sub-contractor, for example instructing 'free' opening up or testing;
- disputes relating to extensions of time.

Perhaps the most important exception to the embargo is that relating to payment, as the fiercest disputes, whatever their roots may be, usually turn upon the withholding of money from the sub-contractor.

Another important characteristic of the JCT arbitration clause is the breadth of the powers it gives to the arbitrator. The most significant of these is his power to 'open up, review and revise any payment, certificate, opinion, decision, requirement or notice'. It is suggested by Neil Jones in his book *A User's Guide to the Arbitration Rules* that this allows the arbitrator to substitute his own discretion for that of the architect where the architect is exercising professional judgment, as opposed to where he is acting as agent of the employer. For example, in deciding what variations to order, the architect is acting solely as the employer's agent, whereas in certifying the value of that varied work he has a duty to be fair to both sides.

Whilst it is submitted that that view is correct, it could give rise to considerable difficulties in the context of sub-contracts, where it is often the main contractor rather than the architect who takes the decisions on such vital matters as practical completion and extensions of time. The rules at main contract level could encourage a party to contend that the main contractor's role cannot properly be compared to the certifying role of the architect, and therefore that his decisions cannot be revised by an arbitrator. However, it is suggested that such an argument would not succeed, because it is not supported by the wording of the arbitration clause in the sub-contracts, which does not limit the certificates, decisions, notices etc. which can be opened up to those issued by the architect, but leaves the matter open.

It was confirmed by the courts in *Northern Regional Health Authority v. Derek Crouch Construction Co.* [1984] QB 644 that a court determining a dispute under a JCT form of contract did not have such wide powers. Browne Wilkinson LJ summarised the position very clearly when he said:

> The limit of the court's jurisdiction would be to declare inoperative any certificate or opinion given by the architect if the architect had no power to give such certificate or opinion or had otherwise erred in law. The court

could not (as the arbitrator could) substitute its discretion for that of the architect.'

This principle has now been modified by Section 100 of the Courts and Legal Services Act 1990, which provides that the court may exercise specific powers conferred on the arbitrator by an arbitration clause if the parties to the arbitration agreement so agree. However, it will frequently be in the interests of one party to withhold that agreement, and so it remains to be seen how effective Section 100 will be.

In addition to his powers in relation to certificates, decisions and notices, the arbitrator is given express powers to rectify the contract (in other words to amend it to reflect the true intention of the parties) and to direct such measurements or valuations as are in his opinion desirable to determine the rights of the parties.

Joinder

As many sub-contractors know to their cost, the vast majority of disputes between main contractor and employer arise out of or involve the activities of sub-contractors. It therefore makes sense if a dispute between sub-contractor and main contractor, which is closely related to a dispute between main contractor and employer, can be heard at the same time by the same arbitrator. However, arbitrators do not have the inherent powers enjoyed by the courts to join related proceedings, and therefore this can only be achieved if the parties so agree or if there are express provisions for joinder in the relevant arbitration agreements.

All of the JCT main contracts and sub-contracts under consideration here do contain such express provisions, but they are optional. The option will be exercised at main contract level, so the joinder provisions in the sub-contract will apply only if the relevant joinder provisions apply to the main contract.

Taking the joinder provisions in NSC/C as an example, if a dispute under the sub-contract raises issues which are 'substantially the same as or connected with issues raised in a related dispute' between the employer and main contractor under the main contract which has already been referred to arbitration, the dispute under NSC/C will be referred to the same arbitrator, who is expressly given the same powers as the High Court in relation to joinder. However, either the main contractor or the sub-contractor may require the dispute to be referred to a different arbitrator if they reasonably consider the arbitrator already appointed is not qualified to determine the dispute under the sub-contract (although this is difficult to envisage bearing in mind that if a dispute is related it is likely to turn upon similar facts).

The arbitration clause in JCT 80 contains a reciprocal arrangement whereby if the dispute under the sub-contract has already been referred to arbitration, the related dispute under the main contract will be referred to the same arbitrator, subject to the same right to reasonably require the appointment of a different arbitrator on the grounds of inappropriate qualifications.

Name-borrowing arbitrations

Under sub-contracts such as NSC/C and Works Contract/2 some of the sub-contractor's rights and liabilities, including his entitlement to payment, will depend upon certificates issued by the architect under the main contract. Because the architect is not a party to the sub-contract, if he fails to act properly or at all, the sub-contractor would not normally be able to take action against him. NSC/C and Works Contract/2 overcome this problem by permitting the sub-contractor to commence arbitration proceedings under the main contract in the main contractor's name against the employer for a breach of contract by his agent, the architect. This right is subject to the sub-contractor giving the main contractor such indemnity and security as he may reasonably require.

Although this arrangement sounds simple enough, it can cause considerable practical problems if the main contractor has an interest in the outcome of the name-borrowing arbitration. For example, in *Lorne Stewart plc* v. *William Sindall plc* (1986) 35 BLR 109 the main contractor did not wish actively to participate in a name-borrowing arbitration because his own arbitration against the employer had been settled. The court held that the following term was implied into the sub-contract:

> 'Upon the sub-contractor exercising its right to use the main contractor's name in an arbitration against the employer, the main contractor will render to the sub-contractor such assistance and co-operation as may be necessary in order to enable the sub-contractor properly to conduct the said arbitration.'

This implied term should be of considerable assistance to sub-contractors in future name-borrowing arbitrations.

The JCT Arbitration Rules

In July 1988 the JCT published a set of Arbitration Rules, and all of the standard JCT forms and their related sub-contracts now provide that arbitrations arising under those contracts must be conducted in accordance with the Rules.

The Rules provide for three different types of procedure:

(i)   procedure without a hearing, where the case will be decided on the basis of written evidence only – Rule 5;

(ii) full procedure with hearing which provides for, in effect, a full set of pleadings and discovery of documents, as well as an oral hearing – Rule 6;

(iii)   short procedure with a hearing, where only a simple written statement identifying the matters in dispute will be prepared, and the arbitrator will reach his decision primarily on the basis of oral submissions by the parties – Rule 7.

The parties to the dispute are entitled to agree which procedure should apply, but if they will not or cannot agree, the arbitrator must direct that either Rule 5 or Rule 6 will apply. Note that the arbitrator cannot require the arbitration to be conducted under Rule 7; the short procedure with a hearing is applicable only with the agreement of the parties.

The Rule 5 procedure, without a hearing, will be suitable for cases which do not involve disputes on the facts requiring oral evidence and cross-examination. A dispute as to the construction of the contract, for example on whether a requirement in the specification conflicts with a condition of the contract, would be particularly suitable for this procedure.

The Rule 7 procedure will be appropriate where a quick, binding decision is required to allow the parties to get on with the works with the minimum delay – for example where there is a dispute as to whether an item of equipment complies with the specification, or whether an instruction to carry out further testing following the discovery of an initial defect is reasonable. The hearing can be held on site, if appropriate, so that the arbitrator has an opportunity to inspect the works.

Under Rule 7.1.2, each party bears their own costs of a Rule 7 arbitration unless the arbitrator 'for special reasons' directs otherwise. The rules do not specify what those special reasons might be, but it is thought that if one side is pursuing a thoroughly unmeritorious argument with the assistance of lawyers, the other side may recover their costs if they engage lawyers to rebut that argument. The rule should, however, discourage the use of legal or other professional representation unless it is absolutely necessary, and thus ensure that Rule 7 arbitrations remain fairly informal.

The full procedure in Rule 6 will be appropriate for all other disputes, particularly those where detailed oral evidence is required to establish the facts.

An excellent and detailed commentary on the JCT Arbitration Rules is given in Neil F. Jones' book referred to on page 231; readers who require further information on the Rules or on arbitration generally are referred to that text.

Appeals from an arbitrator's award

As the law on arbitration has developed, it has become progressively more difficult to appeal from an arbitrator's award. The Arbitration Act 1979, for example, severely restricts the right to appeal to the courts on a point of law.

There are two stages at which a point of law may arise: during the course of the arbitration (i.e. before the arbitrator has given his decision), and out of the arbitrator's award. If a question of law arises during the course of the arbitration, the parties may only apply to the court for a decision on it with the arbitrator's consent and the leave of the court, or with the consent of all the parties. If a question of law arises out of the arbitrator's award, an appeal can be brought to the court only with the consent of the parties or the leave of the court.

Section 1(4) of the Act provides that the High Court should not grant leave to appeal on a question of law arising out of an arbitrator's award unless it considers that, having regard to all the circumstances, the determination of that question of law could substantially affect the rights of one or more of the parties to the arbitration agreement. Similar considerations should be taken into account when deciding whether to allow a question of law arising during the course of the arbitration to be determined by the courts, but in this case the court must also be satisfied that it might also produce substantial savings in costs.

The Courts exercise their discretion to grant leave sparingly, applying both the rules summarised above and the guidelines set down by the House of Lords in *Pioneer Shipping* v. *B.T.P. Oxide (The Nema)* [1982] AC 274. In *The Nema*, it was held that in cases concerning a 'one-off' contractual provision, leave should not normally be given unless it was apparent from simply reading the reasoned award that the arbitrator was obviously wrong in the way in which he interpreted the clause. If the construction of a term in a standard form of contract is at issue, leave should be given more readily, as it is in the public interest for the meaning of such forms to be as settled as possible. However, even in these circumstances, leave should not be given unless there is a strong prima facie case that the arbitrator was wrong.

In addition to its power to grant leave to appeal from an arbitrator's award on a point of law, the court has a limited power under the rules of the Supreme Court to remit a matter referred to them for reconsideration by the arbitrator. This power may be exercised where, for example, the arbitrator has made an error which amounts to misconduct but does not justify his removal, or where he admits a mistake.

Litigation

Despite the extensive provisions on arbitration in the standard forms, many building contract disputes find their way into the courts. This is because there are a number of ways in which an arbitration clause can be circumvented, for example:

- by agreement between the parties;
- by proving that there is no 'dispute or difference';

- by issuing a writ – if the other side then takes a 'step' in the proceedings, such as by serving a defence, he cannot subsequently apply to have the court proceedings stayed in favour of arbitration.

A sub-contractor anxious to have his case heard by the court may argue that there is no 'dispute or difference' in cases of clear breach of contract by the main contractor such as failure to follow the rules regulating set-off, or wrongful deduction of cash discount. Although such arguments have succeeded in the past (notably in cases such as *Ellis Mechanical Services* v. *Wates Construction* (1976) 2 BLR 60 and *Chatbrown Ltd* v. *Alfred McAlpine Construction (Southern) Ltd* (1986) 35 BLR 44) in recent years the courts have become much more reluctant to get involved in disputes where the contract contains an arbitration clause. Perhaps the case of *Hayter* v. *Nelson* (1990) 2 Lloyd's Rep 265 is the high watermark of this trend, when a judge commented that even though one party might be indisputably right and the other indisputably wrong, this did not mean that there was not a dispute which fell within the ambit of the arbitration clause.

One of the reasons why a plaintiff (i.e. the person bringing the case) may prefer litigation is the availability of summary procedures which can produce a decision relatively quickly and for a relatively modest outlay in legal fees. The most important summary procedures in relation to building contracts are .described below.

Summary judgment under RSC Order 14

The plaintiff may apply for summary judgment if he believes that there is no defence to his claim. The application is made after the plaintiff has issued a writ and served a statement of claim (which is the pleading setting out the basis of his case).

It is relatively easy for the defendant to defeat an application for summary judgment, as he only needs to prove that he has an arguable defence to the plaintiff's claim.

Interim payment under Order 29

As an alternative to applying for summary judgment, or in addition to it, a plaintiff who feels he has a strong case may apply to the court to order the defendant to make an interim payment to him. The court has such a power where:

- the defendant has admitted liability but there is still a dispute as to the amount of damages he is liable to pay; *or*
- judgment has been given for the plaintiff, but the damages are still to be assessed; *or*
- the court is satisfied that, if the action proceeded to trial, the plaintiff would be given judgment for a substantial amount.

The majority of the case law on Order 29 concerns the third of these possibilities. In *Shanning International Ltd* v. *George Wimpey Ltd* [1988] 43 BLR 36 it was held that, when deciding whether to award an interim payment where there is no admission or judgment as to liability, the court should apply a two stage test. Firstly, it must be satisfied that if the action proceeded to trial, the plaintiff would obtain judgment for a substantial sum. In reaching that decision, any set-off or counterclaim which the defendant may have must be taken into account. Only if the court is so satisfied can it go on to decide whether the discretion to order an interim payment should be exercised.

The burden of proof on the plaintiff in an application under Order 29 is a heavy one; in *Crown House Engineering Ltd* v. *Amec Projects Ltd* (1989) 48 BLR 32 Bingham LJ said that it permitted the court to order an interim payment 'to the extent that a claim, although not actually admitted, can scarcely be effectively denied'.

Procedure in High Court actions

A full description of court procedure is beyond the scope of this book, but the basic steps leading up to a trial are summarised below.

An action is begun by the issue of a writ by the person who wishes to claim damages or some other relief, such as an injunction. Once a writ has been issued, the relevant limitation period stops running, although the writ need not be served until four months after the date of issue.

The person to whom the writ is addressed (the defendant) must acknowledge service of it within fourteen days of the date of service. Writs should never be ignored, because if the defendant fails to acknowledge service within the relevant period, the plaintiff can apply for judgment against him (a default judgment).

Documents setting out each party's case known as pleadings are then exchanged. After the close of pleadings discovery takes place. This involves the listing and inspection of all of the documents in each party's possession which are relevant to the case. In construction cases these are usually voluminous, which means that discovery is a tedious and expensive part of the process.

After the exchange of pleadings there is a preliminary hearing called a summons for directions, at which the judge will set out a timetable for the remainder of the preparation for trial, and set a date for the trial itself.

Official referees

Many of the major cases arising out building contracts are tried by Official Referees. Official Referees, or 'ORs' as they are often known, are judges nominated by the Lord Chancellor to deal with 'Official Referees' Business' which is certain specialist cases including those related to building contracts.

Because of their considerable experience in hearing cases concerning construction law, ORs are well versed in the practices of the industry and the standard forms of contract. They have also developed a number of mechanisms which are particularly useful in building cases. These include the 'Scott Schedule', which is a document in tabular form setting out the views of both sides on certain aspects of the dispute, such as allegations of defective work.

Once a case has been allocated to a particular OR, he will hear all of the various interlocutory applications as well as the trial itself. This enables him to exercise a much greater degree of control over the preparation for the trial than the average High Court judge, and for a time resulted in relatively quick decisions in cases which went before the ORs. However, in the early 1990s the ORs became, to an extent, victims of their own success, and the sheer number and complexity of cases with which they were required to deal meant there was often a considerable delay before trial.

The procedure in the ORs courts is similar to that in the High Court, save that an application for a summons for directions should be made within 14 days of the defendant giving notice of his intention to defend the claim, rather than after the close of pleadings.

County Courts

Smaller cases are heard in the County Courts. Their jurisdiction has been dramatically increased by the Courts and Legal Services Act 1990: they can now try cases where the sum in dispute does not exceed £50 000, and if it is less than £25 000, the plaintiff must start the action in the County Court. The procedure in the County Courts is broadly similar to that in the High Court, although there are no specialist judges such as ORs available to try construction cases.

Arbitration and litigation compared

Litigation and arbitration are, potentially, enormously expensive and time-consuming undertakings. They almost invariably involve teams of lawyers on both sides, and months if not years of work from all those who were connected with the contract. In this they are virtually identical. There are, however, certain points of distinction between the two procedures which anyone with the choice of issuing a writ or a notice of arbitration would do well to bear in mind. For example:

Jurisdiction

An arbitrator appointed under a JCT standard form or one of its related sub-contracts has the power to open up, review or revise certificates, opinions, decisions, etc. issued under the contract, whereas a judge will only have that power if *both* parties choose to give it to him.

Privacy

Arbitrations are held in private, whereas court proceedings are public. This can be a crucially important factor if the dispute is potentially embarrassing for one of the parties.

Multiplicity of parties

Joinder of parties and third party proceedings are a relatively simple matter in the courts, but can be problematic in arbitration. Although the joinder provisions in the JCT forms go some way towards overcoming this problem, they are optional and therefore will not be included in every contract, and they will be of no help if the dispute involves a sub-contractor employed on a non-standard form or a member of the professional team.

Time

Whether either procedure offers a speedier solution will depend on the circumstances of the case. The summary procedures described above can mean that the courts offer the quickest remedy, but this will depend upon the strength of the plaintiff's case. If the JCT arbitration rules apply, one of the short procedures may result in an earlier decision than would be available through the courts.

Costs

Costs are closely linked with time. Generally speaking, the quicker a dispute is resolved, the lower the costs.

Arbitration does involve some additional cost items such as paying for the hire of premises for the arbitration and paying the arbitrator's fees, but against this must be weighed the costs of legal representation. Only lawyers have rights of audience in the courts, whereas in an arbitration the parties can be represented by whoever they choose. Claims consultants who specialise in arbitrations tend to charge on a similar basis to solicitors and barristers but, overall, their fees may be lower.

Effect on business relationships

Neither litigation nor arbitration is likely to enhance a business relationship, but some parties feel that arbitration is less inherently aggressive, perhaps because it is held in private, and therefore feel it is the less damaging of the two.

Adjudication

Adjudication is a contractual mechanism for determining a dispute by refer-ring it to a third party named in or appointed under the contract who will give a quick decision, usually based only on written evidence, which will bind the parties until practical completion.

Adjudication is often described as a type of ADR (alternative dispute resolution), and whilst this is correct in the sense that it provides an alter-native to arbitration and litigation, there is one very important difference. Adjudication involves the imposition of a decision by the adjudicator on the parties, whereas 'true' ADR results in the parties reaching their own settlement with the assistance of a neutral adviser. Although that assistance may include giving an opinion on the merits of the dispute, the parties are not bound by that opinion, and may ignore it completely if they wish to do so.

The standard forms provide for adjudication in two instances: in the sub-contracts in relation to disputes on set off and in the Design and Build Form 1981 on certain 'Adjudication Matters'.

Adjudication under the sub-contracts

The adjudication clauses in the various standard forms are in virtually identical terms, so no distinction is drawn between them here.

The identity of the adjudicator should be agreed by the parties to the sub-contract at tender stage, and his name inserted in the sub-contract appendix (or in the case of forms such as NSC/C in Part 3 of the tender document NSC/T). This is often not done, and the relevant entry is either left blank or 'to be agreed' is inserted. In 1987 the sub-contracts were amended to cater for this eventuality, and now they provide that if no adjudicator has been named, or if he is unable or unwilling to act, the sub-contractor may unilaterally appoint an adjudicator from the list maintained by the Building Employers Confederation. The BEC does not actually propose an adjudicator to the sub-contractor, it simply makes the list available to him from which he makes his own selection.

If the sub-contractor wishes to refer a dispute on set-off to adjudication, the procedure is as follows:

- within 14 days of receiving the main contractor's notice that he intends to make a set-off, the sub-contractor sends a statement by registered post or recorded delivery to the main contractor setting out his reasons for disagreeing with the set-off and the particulars of any counterclaim (for example if the set-off relates to delay and the sub-contractor alleges he is entitled to an extension of time, his counterclaim might be for loss and expense);

- at the same time as he issues that statement, the sub-contractor must request the adjudicator to act and issue a notice of arbitration (although few disputes submitted to an adjudicator actually end up in arbitration);
- within 14 days of receiving the sub-contractor's statement, the main contractor may send a statement to the adjudicator by registered post or recorded delivery giving brief particulars of his defence to any counterclaim (but he has no right of reply to the sub-contractor's statement regarding the set-off itself, and therefore will be judged solely on his original notice of set-off);
- the adjudicator must give his decision within 7 days of receiving the defence to counterclaim or, in the absence of such a defence, within 21 days of receiving the original request from the sub-contractor.

The adjudicator can either direct that the amount set off should be retained by the main contractor, or that it should be paid to the sub-contractor, or that it should be paid to the trustee stakeholder (a bank named in the appendix or selected by the adjudicator) pending agreement or arbitration. He may also adopt any combination of these three options, for example by ordering that part of the money should be retained by the main contractor and the rest paid to the trustee stakeholder.

How an adjudicator's award should be enforced was recently considered by the courts in *A. Cameron Ltd v. John Mowlem & Co. plc* (1990) 52 BLR 42. In that case, a sub-contractor employed on the terms of DOM/1 obtained an award in their favour from an adjudicator, but the main contractor failed to comply with the adjudicator's decision. Cameron made two attempts to enforce the adjudicator's award through the courts: they applied to the court to enforce it under Section 26 of the Arbitration Act 1950 as if it were an arbitrator's award, and they applied for summary judgment under Order 14 on the basis that the main contractor was in breach of contract by failing to comply.

The Court of Appeal held that an adjudicator's decision should not be treated as if it were an arbitrator's award because it was not sufficiently finite, and therefore refused to enforce it under Section 26. This aspect of the decision came as no surprise to most commentators, and in no way detracts from the validity of adjudication.

Perhaps more surprisingly, the court also refused to grant summary judgment, on the grounds that, although the sub-contract expressly provided for payment to be made to the sub-contractor immediately on receipt of an adjudicator's award to that effect, the main contractor need not pay any greater sum than was due under the payment provisions. Mowlem had argued that, regardless of any claim they may have to set off loss and expense, no sums were due to the sub-contractor because of disputes relating to the valuation of the sub-contractor's work, relating both to quality and the amount of work executed.

The Court of Appeal confirmed that the adjudicator's jurisdiction did not extend to such matters, and that disputes relating to valuation of the amount due under the sub-contract (as opposed to disputes relating to the deduction of loss and expense) could only be resolved by agreement between the parties or by arbitration.

In a later case, *Drake & Scull Engineering Ltd* v. *McLaughlin & Harvey plc* (1992) 60 BLR 102, the court did issue an injunction requiring the main contractor to comply with an adjudicator's award. The facts in that case were slightly different from *Cameron* v. *Mowlem*, as the main contractor had argued that the adjudicator had exceeded his powers by ordering the whole of a particular sum payable under the sub-contract to be paid to the trustee stakeholder, not just that part of the set-off disputed by the sub-contractor.

Adjudication is a useful mechanism for sub-contractors to challenge a wrongful set-off in that it is quick and relatively cheap. However, as *Cameron* v. *Mowlem* demonstrates, there may be problems in enforcing the adjudicator's award if there is any dispute as to the valuation of the sub-contractor's work, and if the adjudicator orders the sum in dispute to be paid to the trustee-stakeholder it will be of only minimal comfort to the sub-contractor, as it will not actually help his cashflow.

*Adjudication under the JCT design and build form

Adjudication will only be an option under JCT 81 where it is stated in the appendix that the Supplementary Provisions apply. Where this is the case, the mechanism has a much wider application than under the sub-contracts, as all disputes arising before practical completion on what are described as the 'Adjudication Matters' are referred to adjudication rather than arbitration. The Adjudication Matters include alterations or adjustments to the contract sum (i.e. variations and fluctuations), whether or not work is defective and extensions of time.

The adjudicator should be named in the appendix to the contract, but if no name has been inserted or he is dead, an adjudicator can be appointed by the person selected as appointer of an arbitrator. If a person has been named but is unable or unwilling to act, he may appoint an appropriately qualified person in his stead.

The process itself is an extremely simple one. Either party may give the other a written notice that they wish to refer a dispute to adjudication. Not later than 14 days after the date of that notice, both parties send the adjudicator a written statement setting out the matters in dispute. Within 14 days of receipt of those statements, the adjudicator tells the parties when he expects to be able to give a decision. If he wishes, he can require them to provide further documents or information to enable him to understand the dispute more fully.

The decision reached by the adjudicator becomes an 'Adjudicated Provision' of the contract with which the parties must comply, and which takes precedence over any conflicting clauses in the contract. If the parties wish to challenge the decision, they may do so in arbitration after practical completion.

There is no case law on enforcing an adjudicator's decision under JCT 81, although *Cameron* v. *Mowlem* indicates that an application for summary judgment would probably be the best strategy.

Alternative dispute resolution

The vast majority of disputes where arbitration or litigation has been commenced are settled before there can be a full trial before a judge or arbitrator. This is sometimes because preparation for a hearing results in a clarification of the issues, but the scale of the legal fees involved in bringing a case to trial is much more likely to be the factor which drives the parties towards a compromise.

Until recently such settlements were reached on an informal basis, usually through direct negotiations by the parties themselves, or by their lawyers or other professional advisers. In other jurisdictions, notably the United States, more sophisticated methods of reaching a negotiated settlement have recently emerged, usually involving the intervention of a skilled neutral adviser who helps the parties to resolve their dispute. These methods of alternative dispute resolution, or ADR, are now becoming available in the UK.

There is a whole range of ADR techniques, but those most likely to be used in the building industry are mediation, executive tribunals and non-binding adjudication. A key characteristic of ADR is flexibility of procedure, which makes it difficult to generalise about the role of the neutral adviser or the way in which any ADR process is likely to be conducted. However, the broad principles set out below are followed by many practitioners of ADR, and are based on the NJCC's Guidance Note on the subject.

Mediation

Mediation is the most widely used method of ADR, and already has a good track record in the settlement of disputes in other fields, such as industrial disputes (through ACAS) and family law.

In a mediation, the parties are helped to resolve their dispute by a neutral mediator, who should be an independent person skilled in both the subject matter of the dispute and in mediation techniques. The mediator's task is to facilitate communication, to identify the interests of each party (which may be quite different from their strict legal rights) and to help them to find ways of solving their problems. If that is the full extent of the mediator's role, the mediation is described as 'facilitative'. In some cases, the parties will want the mediator to go a step further and point out the strengths and weaknesses of their respective cases, in which case the mediation is described as 'evaluative'.

However, even in an evaluative mediation, the mediator does not decide the dispute between the parties and does not impose a decision upon them. Although he may suggest terms of settlement, they will not be binding unless the parties choose to accept them and to enshrine their agreement in a legally binding document.

The parties and the mediator will decide how the mediation is to be conducted, often at a preliminary meeting before the day of the mediation. A typical mediation might go through the following stages:

- each party produces a brief written statement of its case and submits it to the mediator in advance — mediators often stipulate a maximum length for such submissions to ensure that the parties (and their lawyers) concentrate on the key issues;
- on the day of the mediation, each party presents an oral summary of its views to the mediator in the presence of the other party;
- the parties then go into separate rooms, and the mediator talks to each of them to try to identify their real interests (these are often described as caucuses);
- the mediator may enter into 'shuttle negotiations' in which he suggests to one party settlement terms which have either been put forward by the other party or which he thinks might be acceptable to them;
- if an agreement is reached, the parties come together for a final session with the mediator to draw up and execute an agreement which, if properly drafted and executed, would be enforceable in the courts.

Executive tribunal

An executive tribunal tends to be more formal than a mediation. The tribunal is a panel made up of a senior representative of each party who has not been involved in the day-to-day conduct of the dispute, typically the chief executive, together with a neutral adviser.

Preparation before the 'hearing' may include limited discovery of documents, and it is more likely that the parties' lawyers will be involved than in a mediation.

On the day of the tribunal, each party's representatives (often lawyers although there is no requirement for legal or other professional representation) will make a short presentation to the panel setting out the key elements of their case. The panel then withdraws and, with the assistance of the neutral adviser, attempts to reach a settlement. At this stage, the process often becomes similar to a mediation.

If terms of settlement can be agreed, they will be set out in a formal agreement which is executed by the parties.

Non-binding adjudication

Often the parties to a dispute are anxious for the opinion of a neutral third party as to the strength of their case, or they may be genuinely in dispute as to the law on a particular issue, or as to how a certain contract condition should be interpreted. In all of these cases, non-binding adjudication may provide the answer.

The process involves each party making an oral or written presentation to an independent individual, who gives his or her views on the merits of the dispute. Although that opinion will not be binding, provided both sides trust the skill and judgement of the adjudicator, it will often be enough to persuade them to reach a compromise.

When to use ADR

Whenever a dispute arises, the parties can attempt to settle it through ADR. It is not necessary for the contract between the parties to contain a clause requiring or contemplating ADR, although such clauses can be useful as they reduce the risk that suggesting ADR will be seen as a sign of weakness by the other side.

If ADR is to have any real chance of success, it is vital that those involved in the process have the authority to bind the party they represent. This will often dictate the timing of ADR, as very senior people may not become involved until the dispute has escalated beyond the norm, or involves particularly large sums of money.

The savings in time and cost which will be made if a settlement results from the ADR process will be greatest if it is initiated before formal litigation or arbitration has been started. Unfortunately, it will not always be a viable option at such an early stage, as the only way to persuade the other party to the dispute to consider settlement may be to issue a writ or a notice of arbitration.

Lawyers often argue that it is not until a full investigation of the facts has taken place — usually after discovery — that a realistic settlement can be achieved. This reveals a fundamental misunderstanding of ADR, which aims to reach a compromise based on the interests of the parties, rather than on their strict legal rights.

If litigation or arbitration has already been commenced when ADR is attempted, it is usual to request a stay of proceedings while it takes place.

When not to use ADR

Despite what some of its supporters claim, ADR is not a cure-all, and in some circumstances may do more harm than good. The Centre for Dispute Resolution (CEDR), which is a non-profit making body set up to promote ADR in the UK, advises that ADR will not be appropriate if:

- the parties want a legal precedent out of the case;
- they want the case to be heard in public;
- an injunction is urgently required to preserve rights or property;
- auditors or others require a decision imposed by a third party;
- the dispute is genuinely capable of being solved by direct negotiation;
- the other party to the dispute has no interest in settlement (although this category of dispute should be treated with care, as in many cases both sides claim it is the other who does not wish to settle).

Advantages of ADR

If it works, ADR undoubtedly has a number of advantages over arbitration and litigation. Apart from saving time and money, the process is less inherently adversarial than formal legal proceedings, and so commercial relationships can be protected.

An important characteristic of an ADR settlement is that it can encompass matters over which a judge or arbitrator has no jurisdiction. The courts can only grant injunctions or award damages, and although arbitrators have slightly wider powers, they are also limited. In contrast, an ADR settlement can include whatever the parties want it to include, such as agreements relating to future contracts.

ADR is also highly flexible – it is not regulated by statute or other procedural rules, and although this can be unsettling for lawyers who are unfamiliar with the process it allows the parties to adopt any method of negotiation which is likely to result in a settlement.

Disadvantages of ADR

ADR has two important drawbacks, which should not be underestimated.

Firstly, there is no guarantee that it will resolve the dispute because, unlike a judge or arbitrator, the neutral adviser cannot impose a solution on the parties. For all their drawbacks, litigation and arbitration produce 'an answer', whereas if the parties to an ADR process fail to reach agreement they will have to revert to more conventional methods of resolving the dispute. If this happens, ADR can actually increase costs and cause delay, although it may clarify the issues and result in settlement earlier than it might otherwise have been achieved.

Secondly, ADR presupposes that both sides either want or need to resolve their dispute. It is therefore unlikely to get off the ground if it involves the flexing of commercial muscle by, for example, simply withholding payment. This is perhaps the most serious disadvantage in the context of sub-contracting, when the commercially weaker party (i.e. the sub-contractor) will usually be the pursuer rather than the pursued.

In addition, those contemplating ADR are often concerned about the effect any disclosures they may make during the process will have on their case if the ADR fails and the dispute eventually goes to arbitration or the courts. To an extent, the parties can control this risk, as they will be able to control the amount of information they reveal to the mediator and to the other side. However, if the negotiations are to have any realistic prospect of success, a certain amount of frankness, to the mediator if not to the other side, will be essential.

The issue of confidentiality should be dealt with in the mediation agreement, which should provide that the whole process will be on a 'without prejudice' basis, which means that no documents which come into being for the purpose of the process can be used in any subsequent arbitration or litigation. The same rule applies to any statement made in the course of the ADR process. The parties also usually agree that the mediator may not be called as a witness by either side.

Finally, it should be borne in mind that ADR is a relatively untried process in the UK, and that although it can undoubtedly achieve great savings if successful there is, as yet, little empirical evidence that it has been widely adopted.

Index